JOHN JOLY

RADIOACTIVITY AND GEOLOGY

AN ACCOUNT
OF
THE INFLUENCE OF RADIOACTIVE ENERGY
ON TERRESTRIAL HISTORY

Elibron Classics
www.elibron.com

Elibron Classics series.

ISBN 1-4021-3577-7 (paperback)
ISBN 1-4021-0106-6 (hardcover)

This Elibron Classics Replica Edition is an unabridged facsimile of the edition published in 1909 by Archibald Constable & Co., Ltd., London.

RADIOACTIVITY AND GEOLOGY.

RADIOACTIVITY AND GEOLOGY

AN ACCOUNT

OF

THE INFLUENCE OF RADIOACTIVE ENERGY
ON TERRESTRIAL HISTORY.

BY

J. JOLY, M.A., Sc.D., F.R.S.,

PROFESSOR OF GEOLOGY AND MINERALOGY IN THE UNIVERSITY OF DUBLIN.

LONDON:
ARCHIBALD CONSTABLE & CO., LTD.
1909.

DUBLIN:
PRINTED AT THE UNIVERSITY PRESS,
BY PONSONBY AND GIBBS.

PREFACE.

SHORTLY after the meeting of the British Association in Dublin last year, it was suggested to me by several friends that my Presidential Address to Section C might with advantage be re-written in a more extended and less technical form. The same suggestion reached me from a few well-known publishers, foremost among whom was the firm whose name appears on the title-page. The idea was already in my mind, for I had been compelled, by the limits necessarily imposed on a Presidential Address, to leave unsaid much that I thought should be said, not only for purposes of elucidation, but by way of addition. Thus it was that the present volume originated.

But when the task of re-writing the Address was begun, it soon became apparent that mere expansion of the matter already in print was insufficient to meet the requirements of those who, for various reasons, may be unable to abstract from a highly specialized and voluminous literature the principles underlying this new branch of geological research. It seemed also advisable that, in addition to a simply worded account of the principles involved, some practical details should be given which might aid in the prosecution of further research. The two introductory chapters and the final chapter were written with these considerations in view.

The leading topics dealt with in my Address are, of course, specially dwelt on; and, it is hoped, to their better understanding. As it is impossible to write upon these many-sided questions without new aspects of them being presented, I have added a considerable amount of fresh matter. The chapter on the part which radioactivity may have played in the architecture of mountains refers to views incidentally touched upon in a discussion held at the same Dublin Meeting of the British Association. The chapter on the age of the earth requires, I think, no apology, in view of the interest excited in the application of radioactive methods to the measurement of geological time. The short discussion of the radioactivity of the atmosphere is introduced not only because of its direct application to denudative geology, but on account of its bearing on the question of the distribution of thorium in the surface rocks. It is needless to add that if the evidence afforded by the frequently observed presence of thorium emanation in the atmosphere is found to be substantiated by further work upon the rocks, every conclusion reached in this book as to the geological importance of radioactivity will receive additional emphasis, even if some revision of the numerical estimates be entailed.

There are, of course, many, even among scientific readers, who are still unacquainted with the considerable body of fact which enters into the subject of radioactivity as an influence on terrestrial history. Among these the objection is likely to arise that the time is not properly ripe for the appearance of a book like the present. I ask such critics to reconsider their attitude in view of what is herein placed before them. A large number of experiments on the surface-materials of the earth are concordant in revealing the presence of radioactive elements as widely distributed constituents of the crust. There is nothing *à priori* improbable in the wide-spread occurrence of the parent substance: uranium. Chemical investigations of the type of Sandberger's have already shown the unexpectedly general distribution of the less abundant elements. Spectroscopic analysis extends this conclusion. And here, in the application of radioactive methods, we work for the first time with a means of detection many thousands of times more delicate than the spectroscope. The results of the new analysis are, in short, but an extension of the old.

To these laboriously detailed observations upon the rocks the more generalized ones upon the radioactivity of the ocean have to be added. In its waters and its sediments we are presented with the age-long accumulation of materials arising from the denudation of the primary rocks. Thus the oceanic radioactivity testifies to the world-wide presence of uranium in those surface materials which, throughout the whole of geological time, contributed substances to the seas. When, finally, we compare the results on impoverished detrital sediments with those on the igneous rocks, we find their joint evidence in harmony with that yielded by the ocean.

Although this evidence of the prevalence of radioactive substances might fairly be deemed conclusive, other supporting facts may be mentioned. The presence of helium in rocks and minerals of every description has lately been established. This substance is known to be a product of radioactive change. That the helium in the rocks came into existence as the "*α* particles" evolved by transforming atoms, hardly admits of doubt. Now, the greater part of the energy liberated upon radioactive transformation is associated with the high initial velocity of these atoms. The presence of helium everywhere in the rocks, notwithstanding its continual escape into the atmosphere, is, therefore, independent proof, not only of the wide-spread distribution of the radioactive elements, but of their ceaseless evolution of energy.

But if, on testimony so strong, the ubiquity of uranium be conceded, then the influence of radioactive energy on the surface dynamics of the globe becomes a matter of first importance. The efficacy of uranium as an almost eternal source of thermal energy seems to be unquestionable. At the present moment this fact appears as well founded as any which fallible human judgment has unravelled from the complex phenomena of the universe. We need submit the uranium ore to neither chemical nor physical processes. It is sufficient to take it from the rocks, and place it in the calorimeter, when the constant outflow of heat will be apparent (p. 11). In a word, uranium is evidently in the rocks; and its heat-producing property is beyond question. With these fundamental facts before us, why should we regard as premature their presentation as a *vera causa* in geological dynamics?

There is, indeed, much in this book which is admittedly speculative. Theories regarding the interior condition of the earth are, of course, unprovable. Here we can only choose, among many possible issues, those which are most in consonance with the little real knowledge we possess. Again, in dealing with denudative data we find ourselves often so restricted for want of accurate information, that deductions founded upon existing figures must necessarily be open to modification as knowledge advances. As an instance I might refer to a suggested method of investigating, on the basis of denudative data, the supposed existence of former great continents now submerged in the oceans. The matter is referred to in Chapter VI. Now, the value of the conclusions we draw entirely depends upon the reliability of data at present only imperfectly known. Although we are thus left in uncertainty, the inclusion of such considerations is, I think, justified. The development of geology as an exact science can only be along the lines of perfecting geological mensuration; and any prospective applications of such work which have the effect of emphasizing its importance must be tolerated.

I have received kind assistance from so many friends in obtaining materials for investigation that full acknowledgment is here impossible. More especially Sir John Murray, Professor Agassiz, and Sir Henry Holland have contributed materials which without their aid would have been unattainable by me. To the Trustees of the British Museum I am under similar obligations. It is needless to say that the samples of sea-water from all parts of the ocean were only obtained by the kindness of friends willing to undertake the not inconsiderable labour attending their transmission. Rock specimens, too, of much value have come to me from many parts of the world.

A careful revision of the proof-sheets of this book has been made by Mr. Horace H. Poole. Mr. J. R. Cotter has from time to time given me much help of the most valuable kind. The proofs have also been read by Professor H. H. Dixon. To these friends, as well as to the many to whom individual reference is not made, I have pleasure in tendering my thanks.

J. J.

TRINITY COLLEGE, DUBLIN,
February, 1909.

CONTENTS.

LIST OF ILLUSTRATIONS.

* Reproduced from the *Philosophical Magazine* by kind permission of Messrs. Taylor and Francis.

RADIOACTIVITY AND GEOLOGY.

CHAPTER I.

THE FOUNDATIONS OF RADIOACTIVE SCIENCE.

THE FAMILIES OF RADIOACTIVE SUBSTANCES—GENETIC RELATIONS OF RADIUM WITH URANIUM—THE THEORY OF RADIOACTIVE CHANGE—EQUILIBRIUM AMOUNTS OF GENETICALLY RELATED ELEMENTS—DISCOVERY OF IONIUM—POSITION OF ACTINIUM—THERMAL PHENOMENA ATTENDING RADIOACTIVE CHANGE—THE HEAT-CONSTANT OF RADIUM.

NOTHING can better illustrate the eagerness with which new openings to knowledge are pursued in our time than the history of that branch of science which is devoted to radioactivity. First definitely opened up in 1898 by Mme. and M. Curie, when polonium and radium were discovered, to-day we are in possession of established views in contradiction to the cherished tenets of centuries. It is true that the time-honoured view as to the unalterable stability of the atom fails only in a sense, and not owing to any interference on our part; nevertheless it is now one of the most assured facts of science that the atoms of certain elements change in atomic weight, radiating a portion of their mass and giving up a part of their internal energy in the process. Such, along with a multitude of related facts, fundamentally altering and enlarging our ideas of matter, has been the harvest reaped from the one discovery, and gathered in one short decade.

For our present purpose—the presentment of the new science in its geological aspect—certain facts only in the history of this great advance need be reviewed. The starting-point of the geological importance of radioactivity is the discovery in 1903 by Paul Curie and Laborde[1] that radium steadily maintains its temperature above its surroundings. On thermo-dynamical grounds this result might have been reached at an earlier date, and, indeed, in 1901 Rutherford and McClung had calculated the heat equivalent of the ionization effected by uranium, radium, and thorium.[2] The subject is of fundamental importance, and will be discussed later on. It is first necessary to take a general review of the extent to which, according to present knowledge, radioactivity prevails among the elements.

To-day we are acquainted with at least two families of radioactive substances which appear to be separate and distinct: that which is referable to uranium as the parent or originating substance, and that which appears to be derived from thorium.

The thorium family may ultimately prove of importance to geological science. At the present time its general prevalence among the materials of the earth, to any geologically important extent, remains to be proved. The uranium family includes twelve known or

[1] P. Curie and Laborde, C. R., 136, p. 673, 1903.

[2] Rutherford and McClung, *Phil. Trans.*, A, p. 25, 1901.

PLATE II.

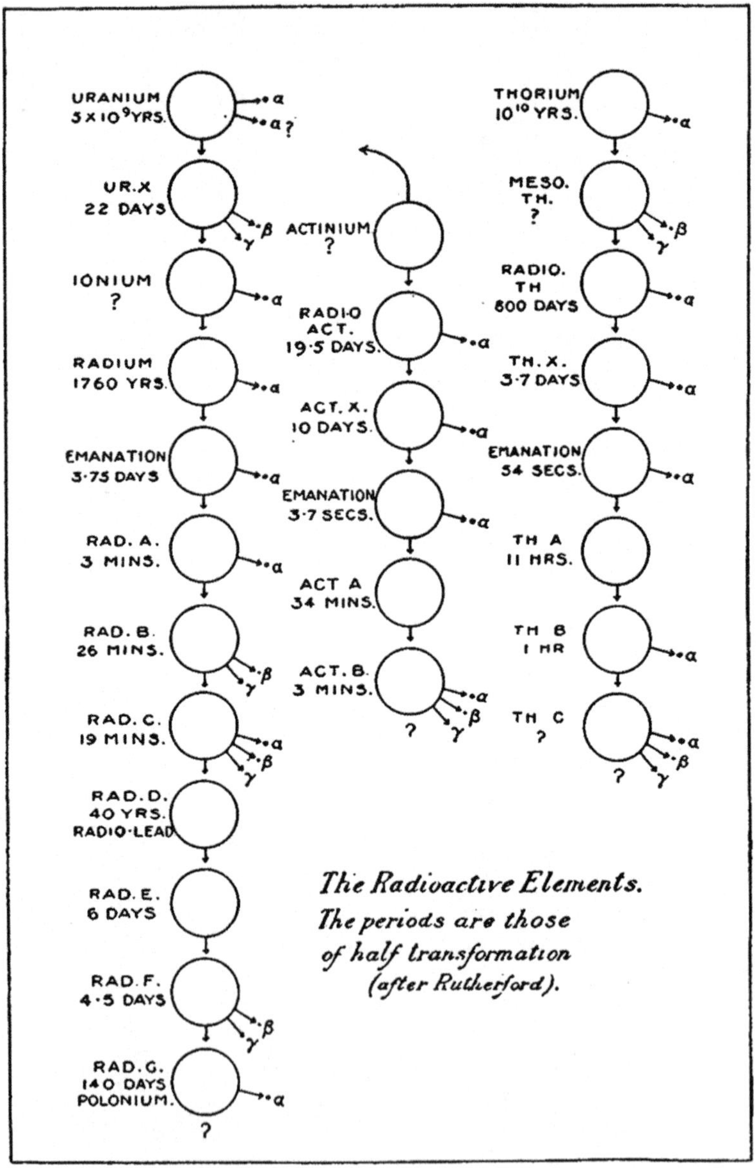

The Radioactive Elements.
The periods are those of half transformation (after Rutherford).

inferred elements in one line of descent from uranium, and probably six other elements derived from uranium on a collateral line. The accompanying diagram, Plate II., after Rutherford,[1] with some modifications to bring it up to recent discoveries, shows the genetic relations of the radioactive elements one to another, as well as the periods of transformation in each case.

It will be seen that radium itself is a comparatively short-lived substance ; its period of transformation—that is, the period required for one-half of any given quantity of the element to transform into the next element in the line of descent—is but 1760 years. This result is obtained by a new method of Rutherford's, whereby the process of radioactive change can be more accurately observed than was hitherto possible.[2] The previous determination by Boltwood afforded 2000 years for this constant.[3] It is therefore ephemeral compared with such periods of time as geology is concerned with; and obviously there must be a progenitor, or source of production, somewhere among the elements.

The view that radium is genetically connected with uranium may be regarded as established. We must consider more closely the grounds for this statement. So far as the geological applications of radium are concerned we might perhaps regard the identification of the originating substance as not of first importance, seeing that *some* such element must exist. Undoubtedly,

[1] Rutherford, *Nature*, March 5, 1908, p. 422.
[2] Rutherford and Geiger, *Proc. R. S.*, A, 81, p. 162.
[3] Boltwood, *Amer. Journ. Science*, June, 1908, p. 493.

however, our knowledge of the parentage of radium strengthens our hands in making these applications; and the quantitative results as regards the total duration of radioactive effects thereby brought into geological science are of the highest value.

The most important evidence for the genetic connexion of radium and uranium is that derived from the examination of ancient minerals.[1] In the hands of Boltwood,[2] and later of Strutt[3] and McCoy,[4] this has revealed the existence of a constant relation between the amount of radium and the amount of uranium present in every uranium-bearing mineral. From various parts of the world—from plutonic rocks, dikes, or metalliferous veins—from whatever source the uranium mineral is derived, the quantitative relation has been found to hold. Now, this remarkable fact is explicable on the theory of the radioactive production of the one from the other, and on this theory only.

To clearly understand this, let us review the fundamental position of the radioactive theory. We can do so in no more interesting connexion.

According to this theory, a radioactive substance is one possessing a certain degree of atomic instability, so that every now and again one of the atoms undergoes partial disruption; the various rays, to which we will later refer, being produced in the act. The instability being something brought about by the

[1] Rutherford and Soddy, *Phil. Mag.*, 65, 561 (1903).
[2] Boltwood, *Nature*, May 25, p. 80, 1904. *Phil. Mag.*, April, 1905.
[3] Strutt, *Nature*, March 17, and July 17, 1904.
[4] McCoy, *Ber. D. Chem. Ges.*, No. 11, p. 2641, 1904.

properties of the atom, is inherent and inevitable; and where many millions of millions of atoms are concerned, on the average a definite fraction of the atoms present at any moment break up per second. Consider now what the result of this must be supposing the newly formed atoms are themselves possessed of an inherent instability, so that in their case also a fractional proportion of the total number present transforms every second. Let us start with the element A. It produces the element B at a rate which depends on how much of A is present, and on the inherent degree of instability possessed by it. B accordingly begins to accumulate. Now, B possesses its own particular degree of instability; and as B accumulates, the number of B atoms breaking up per second increases. At length B accumulates to such an amount that the number of its atoms transforming each second is equal to the number which A is producing per second. Evidently B cannot exist in greater quantity along with the given amount of A, and evidently, also, as A diminishes in amount in course of time, so that its rate of production of B also diminishes, then, simultaneously, the quantity of B which transforms per second must decrease with the diminishing quantity of B. Thus there will still always be the same proportion of B to A. This quantity of B is the amount of B which is in equilibrium with the amount of A present, or the "equilibrium amount." It is a fixed proportion of A, which entirely depends upon the degree of instability possessed by each element—an instability manifested by its rate of change, or "constant of change." In the cases of uranium and radium, the radium is in equilibrium with the amount of uranium present when only a very minute fraction of radium has accumulated. Thus in a gram of uranium the equilibrium amount of radium is only $3{\cdot}4 \times 10^{-7}$ of a gram of radium. No more than this amount of radium can occur as the derivative of the uranium. When ultimately the uranium has diminished to one-half of a gram, then the amount of radium present will be only half the above small quantity, and so on.

The reason of the smallness of this equilibrium amount is the very rapid break-up of the radium compared with the rate of transformation of the uranium. In fact, while radium is transformed at such a rate that half any given quantity of it would be changed in 1760 years, the same would only happen in the case of uranium after the lapse of about 5×10^9 years—five thousand million years.

A hydrostatic analogy may conduce to clearness in the apprehension of this fundamental statement of radioactivity. Suppose a vessel V_1 empties into a vessel V_2 through a narrow tubulure of a certain bore. The bore of the tubulure stands for the atomic conditions controlling the rate of decay or constant of change, say, of uranium. The outflow from V_1 will evidently depend on the amount of liquid contained in it; for the pressure at the orifice of escape is directly as the depth of the fluid. When there is half the fluid, the pressure will be half, and the rate of flow half; and if the depth is again halved, the outflow will again be halved, or be at one quarter its original rate. Thus as time advances in arithmetical progression, the rate of flow diminishes in geometric progression. The quantity of liquid represents the quantity of uranium which thus diminishes in geometrical progression with the arithmetical progression of the time, the final disappearance of the uranium taking a very long time indeed. Now, the rate of flow from V_1 at any instant is, of course, the rate of production of the next substance in the line of descent, which has been called uranium X. (See diagram, Plate II.) If the vessel V_2 has also a tubulure of a certain bore (which may be quite different from that of the first tubulure), there is obviously a certain height to which the fluid will stand in V_2, so that the rate of outflow from it will just equal the rate of inflow from V_1. This height is defined by the rate of inflow from V_1, and the bore of the tubulure to V_2. This bore, again, is the constant of change of uranium X, and the height of the liquid represents the equilibrium amount of uranium X for the particular amount of uranium. If more than this amount of fluid could collect in V_2, the rate of outflow must increase; and as the inflow has remained invariable, the level of the fluid must again be brought back to that of the equilibrium amount. A little consideration will show that if the tubulure of V_2 is wide in bore compared with the tubulure of V_1, the liquid in V_2 will only stand at a small height when its rate of discharge becomes equal to the rate of inflow from V_1. Analogously, the rate of change of uranium X being very rapid compared with that of uranium, the equilibrium amount of the former substance is very small.[1]

The constant of change is indicated by the letter λ, and means the fractional amount of the radioactive substance which transforms each year.

The weight of the evidence afforded by the naturally existing proportionality of uranium to radium in minerals will now be apparent. Both elements are radioactive, and here is a law which can explain the observed proportionality as a direct result of the radioactivity, and a genetic connexion. When it is remembered that the law of radioactive change, to which the fixed proportionality can be traced, has been found to rigorously obtain in all cases which have been quantitatively examined, the evidence becomes overwhelming.

It was thought for a time, before ionium was discovered, that direct observation should show the production of radium from uranium which had been first purified of radium; and when disappointment attended the investigation, the cause had to be sought in the existence of some intermediate substance which, having to take precedence of radium in order of formation, and being slow of change, would retard the appearance of radium. If we imagine on the hydrostatic analogy that the vessel V_3, originally supposed to stand for radium, represents an intermediate substance,

[1] Rutherford, *Nature*, March 5, 1908, p. 422. The author, in an Opening Address to the Dublin University Experimental Science Association, December, 1906, made use of the hydrostatic analogy as above.

and that this vessel V_3 has a very fine bore, it will be evident that a very long time might elapse before any appreciable amount of radium had been received from the vessel V_3. At the time of the experiments which failed to produce radium from uranium, no such intervening element was known. It was plain that the rate of change of uranium X disqualifies it from acting as the retarding substance.

The missing element has recently been found by Boltwood.[1] The new element, ionium, when purified and sealed up, is found to produce radium, and shows a period of transformation comparable to that of radium itself, that is, measured in thousands of years.[2] The failure to obtain radium directly from purified uranium is thus fully explained; and Boltwood considers that ionium is undoubtedly a disintegration product of uranium, intermediate between uranium X and radium.[3] This closes the evidence now in our possession for the genetic connexion of radium and uranium.

Before leaving this part of our subject the position of the actinium group deserves notice. Actinium appears according to Boltwood to be present in uranium minerals in a fixed ratio to the amount of uranium. Rutherford inclines to the view that its connexion with uranium may be explained by supposing that it springs from an indirect line of descent. At some stage

[1] Boltwood, *Amer. Journ. Science*, xxiv., 370, 1907; and xxv., 367, 1908.

[2] Boltwood, *Amer. Journ. Science*, xxv., 378, 1908.

[3] *Loc. cit.*, p. 381.

of descent certain of the transforming atoms, owing to differing states of equilibrium being possible, give rise to a distinct sequence of elements. Such atoms are in subordinate amount, for the activity of actinium present is not such as would be expected if it was in the line of descent of radium.

The experimental evidence of radioactive heating is of the most simple and direct character. The radium salt is enclosed in the Bunsen calorimeter, and the quantity of heat generated in a given time measured directly. Curie and Laborde's first determinations indicated that heat was produced at such a rate that a gram of pure radium must yield sufficient heat in an hour to warm 100 grams of water one degree.[1] This result has been confirmed as being approximately correct by others. Using a differential air-calorimeter, Rutherford and Barnes determined the rate of emission of heat as 110 gram-calories per hour per gram of elemental radium. They moreover showed that this result is due to the successive changes which continually proceed within the radioactive substances: the primary substance, radium, present in the experiment being responsible only for a share of the energy given out.[2]

It is apparent that such measurements as those we have cited are not inclusive of the energy liberated by uranium, uranium X, and ionium, these radioactive substances having been separated from radium in the

[1] P. Curie and Laborde, C. R. 136, p. 673, 1903.
[2] Rutherford and Barnes, *Phil. Mag.*, February, 1904.

chemical process of obtaining the latter element. Direct and careful experiments of the heat given out by Joachimsthal pitchblende have been made by H. H. Poole. The heat-production here includes that of the substances omitted in previous observations. It must also include that given up by the actinium family of radioactive substances. Poole's experiments are not yet concluded, but it would appear that the final results will come out somewhat higher than would have been anticipated. It is noteworthy that in these experiments we have the direct observation of radioactive heating *in situ*, before the chemist has separated the constituents, and in materials which have been subjected to no physical modification at our hands.

Pending the completion and corroboration of these experiments, we proceed on safe grounds by accepting a lesser estimate based directly on the radioactive energy of the several members of the uranium family and on the direct thermal experiments. In a letter received by the author early this present year (1908) Professor Rutherford estimates the rate of evolution of heat involved in the presence of one gram of radium and due to the aggregate energy emitted by the uranium family of radioactive elements, as $5{\cdot}6 \times 10^{-2}$ calories per second. This is probably correct within a few per cent.

CHAPTER II.

PRINCIPLES UNDERLYING THE MEASUREMENT OF RADIUM IN TERRESTRIAL MATERIALS.

THE ESTIMATION OF SMALL QUANTITIES OF RADIOACTIVE MATTER—THEORY OF THE MEASUREMENTS—THE SEVERAL RAYS—IONIZATION OF A GAS BY A RADIOACTIVE BODY—ENERGY OF THE RAYS—THE ELECTROSCOPE AND THEORY OF ITS DISCHARGE BY A RADIOACTIVE SUBSTANCE—NUMBER OF ALPHA RAYS FROM ONE GRAM OF RADIUM—AN ORDINARY MEASUREMENT AND THE NUMBER OF IONS CONCERNED—SENSITIVENESS OF THE METHOD—THE USE OF EMANATION IN RADIUM MEASUREMENTS—CALIBRATION OF THE ELECTROSCOPE—PROPERTIES OF THE EMANATION—EFFECTS OF EMANATION IN THE ELECTROSCOPE—PROCEDURE IN MEASURING RADIUM—PRECAUTIONS—RADIOACTIVITY NOT AFFECTED BY CONCENTRATION—IS IT AFFECTED BY HIGH PRESSURES AND TEMPERATURES?

It occasionally happens that those who have not followed the development of radioactive science express scepticism as to the measurements upon which the vast deductions of radiothermal geology are founded. It appears to them as if quantities so minute as billionths of a gram must be beyond the limits of sufficiently safe determination. This objection only arises from those unacquainted with the mode of measurement, and who therefore possess no mental substitute for the relatively coarse and insensitive methods of chemical science, where the chemical balance, the use of indicators, or even the spectroscope, must mark the limits of perception of small quantities of matter. If science had indeed been restricted to these means of dividing and evaluating matter, it is safe to say that the discovery of radium had never been. But in the application of the electroscope to chemical science—due to the Curies in the earliest days of radioactive investigation—a mode of detection, measurement, and separation was introduced, based upon the radioactive properties themselves, thousands or millions of times as sensitive as the older methods. It must be remembered, however, that the new method applies to radioactive matter only, and may even fail to reveal radioactivity under certain not inconceivable forms of the phenomenon.

A brief account of the new method is necessary here. The method is fundamentally based on the fact that gases become conductors of electricity under the ionizing influence of the rays attending the transformations of radioactive matter. These transformations are accompanied by the emission of :—(*a*) relatively heavy material particles—helium atoms[1]—electrified positively and projected with great velocity. These are the α rays. They represent a definite material part of the atom from which they are derived, and probably indicate that helium enters into the constitution of very many of the elements; (*b*) electrons, or β particles, of very much less mass, but moving at yet higher speeds and electrified negatively; (*c*) the γ radiations, the nature of which is still under discussion, which may be material or ethereal, and which also produce ionization. Of these three forms of radiation the α rays are far the most important, both from the ionization and from the

[1] Rutherford and Geiger, *Proc. R. S.*, 81 A, p. 172; and Rutherford and Royds, *Manchester Memoirs*, vol. liii. (1908), No. 1.

heat-production aspects, the greater importance being due to the more considerable mass which, taken along with the high velocity, involves a larger part of the internal atomic energy than is carried by the other rays. The heat effects traceable to radioactivity are therefore mainly due to the stoppage by surrounding matter of these flying helium atoms, their momentum being spent in producing that atomic *vis viva* which we recognize as heat.

The intimate nature of ionization cannot be said to be known; nor will it be, probably, till we acquire a more minute knowledge of the nature of atoms and atomic systems. It appears most probable that ions are formed in a gas by the detachment from the molecule of a negative corpuscle or β particle, leaving the residual molecule positively electrified. Thus the ions developed in a gas are both plus and minus in electrical sign, two ions originating from each disrupted molecule.[1] There is evidence that the detached β particle does not wander as a free particle, but immediately attracts a surrounding of neutral molecules, so that in effect its mass is greatly augmented. The possession of a free positive or negative charge is a characteristic and essential property of the ion.

The changes brought about in a gas undergoing radioactive ionization may, in mechanical languagé, be described as due to collisions between the projectiles emitted by the radioactive matter and the gaseous molecules, the α ray being, as stated, primarily responsible. It can be shown that many thousands of ions are

[1] See, however, Townsend, *Proc. R. S.*, 80 A, p. 207, and 81 A, p. 464.

created by a single α ray before its energy is sufficiently absorbed to render it incapable of disrupting a molecule. Considerable energy must remain even after this velocity is attained; but no means of observation now at our disposal, save the slow accumulation of helium or, possibly, the development of a sufficient heat effect, would enable us to determine the presence of rays which move with less than the critical velocity required to ionize the particles of a gas or the sensitive molecule of the photographic plate. It is, therefore, possible that certain elements may expel such rays, the fact remaining at present beyond our cognizance. The α rays of radium possess a quantity of energy enormously in excess of that which is associated with the critical velocity; for on an average about 100,000 molecules of gas are ionized by each ray before the ray becomes ineffective.[1] Nevertheless a small thickness of matter in the liquid or solid state suffices to stop the career of one of these rays. Only the thinnest flake of mica will suffer one to pass through. There is little doubt that the microscopic radius of the pleochroic haloes seen in biotite surrounding radioactive particles marks the limits of the penetrability of these rays or of their ionizing influence in the medium.[2]

The nature of the radiations in the case of each radioactive element is given in the diagram Plate II.

The properties of a gas undergoing ionization are easily rendered available for the detection and mea-

[1] According to a recalculation on latest data kindly supplied by Professor Rutherford, February, 1909.

[2] Joly, *Phil. Mag.*, March, 1907, p. 381.

surement of radioactivity. We now pursue the subject with the help of an electroscope, which we easily construct by placing in an enclosed volume of air a small gold leaf, affixed to a strip of brass, supported by an insulating rod of sulphur or quartz, and so arranged that we can readily electrify the leaf. The leaf shows its state of electrification by its repulsion and divergence from the brass strip. Before an ionizing influence is brought to bear upon the surrounding gas there is only a very slow loss of electricity and collapse of the leaf: and, indeed, what there is appears due still to ionization from sources which we cannot control.

Having observed the rate of collapse of the leaf through a low-power microscope with a divided scale in the eyepiece, let us bring up a radioactive substance to the vessel, or, better still, introduce it into the air surrounding the leaf. Immediately the rate of collapse augments and, in fact, may be instantaneous if the radioactivity of the substance is strong enough. To understand this remarkable and important experiment we must consider what goes on in the gas surrounding the electrified leaf. Here the stream of rays coming from the radioactive body is each moment creating ions. True, many of these recombine; but the supply is continually renewed. The ions are, it will be remembered, both plus and minus in sign. Under the influence of the electric charge on the leaf and of the corresponding induced charge upon the walls of the vessel, the positive ions are urged in one direction, the negative in the other, each moving towards the charge of opposite sign and away from the charge of like sign to that which it carries. This constitutes a current, made up of a multitude of minute charges, each ion carrying a fixed and definite amount of electricity. Thus the leaf becomes discharged or electrically neutralized. If the charge upon the leaf is at a sufficient potential, the force upon the ions is adequate to keep them from recombining, urging them at once into the streams of ions and utilizing practically the whole of them as carriers, as fast as they are created by the rays from the radioactive substance. In this case the rate of collapse of the leaf would experience no increase by increasing the voltage or electric pressure: that is, in other words, the rate of discharge is the same whether the leaf shows a wide divergence or a small one. The potential at which all the ions present are utilized as carriers is called the "saturation potential," and the current the "saturation current." Of course the saturation potential varies with the intensity of the ionization. In the case of measurements such as we will further on be referring to, the potential of the gold-leaf system when divergent a few degrees must attain the saturation value.

Before the sensitiveness of the method of measurement which is founded on the above conditions can be fully appreciated, it must be understood that *each* transforming atom of the radioactive substance which emits a ray or rays is potent to affect the conductivity of the gas in the electroscope. We will next inquire into the number of rays from a given quantity of radium, confining our attention to the α rays, which are far the most efficient ionizers.

By a recent brilliant research Rutherford and Geiger have determined the number of rays emitted per second by a known amount of radium. On the simplest and most probable assumption that one atom of radium in breaking up emits one α particle, they find that in one gram of radium $3 \cdot 4 \times 10^{10}$ atoms are transformed each second. But there are three α-ray-producing elements evolved from each of the transformed atoms (exclusive of radium *G*, which takes many years to form); and, according to the experiments of Bragg and Boltwood, these three products emit each an α-ray for one α-ray from the parent radium: so that when the equilibrium amounts of these elements are present, for each α-ray from the radium, there are three rays from its descendants. The total is, therefore, $13 \cdot 6 \times 10^{10}$ rays.[1]

It will be seen later on that in effecting measurements of the radioactivity of terrestrial materials, we do not usually deal with radium itself, but with the first product of change—the emanation which is a gas at ordinary temperatures. This is what we introduce into the electroscope; and we take care to observe the rate of collapse of the leaves after such a period of time that we are assured the successively formed substances, radium A, B, and C, are generated in equilibrium amounts. Hence the α-radiations (see diagram, Plate II.) from three elements have to be considered. Each of these produces $3 \cdot 4 \times 10^{10}$ rays or particles per second, or a total of, say, 10^{11} rays per second.

[1] Rutherford and Geiger, *Proc. R. S.*, 81, A, p. 141.

We can now make the calculation very easily for the case of an actual experiment. We are dealing, let us suppose, with a rock which has only 2×10^{-12} grams of radium in each gram. Ten grams of this rock are brought into solution. The solution is closed in a flask until the emanation is in equilibrium with the radium present. The emanation is then extracted by boiling, and introduced into the electroscope. Now the emanation from one gram of radium gives rise in round numbers, as we have seen, to 10^{11} α-rays per second; but here we have only 20×10^{-12} grams of radium. If n is the number of rays given out by the emanation from so small a quantity of radium, we have:—

$$n : 10^{11} : : 20 \times 10^{-12} : 1,$$

from which $n = 2$. That is two α-rays are liberated by this quantity per second. This small number of rays gives rise to probably *not less* than some 150,000 ions per second, allowing that only about three-fourths of the effect of the rays results in ionization of the air of the electroscope. In electroscopes of the dimensions described further on, this constant production of ions is attended with a quickening of the normal rate of discharge which is unmistakable. Thus the ordinary unavoidable ionization of air in closed vessels may vary from about 10 to 60 ions produced per cubic centimetre per second, or from 4,000 to 24,000 in an electroscope holding 400 cubic centimetres of air[1]; here the number due to the emanation is about 150,000: or the ordinary ionization even at its extreme value is increased more

[1] Rutherford, *Radioactivity*, 2nd ed., p. 534.

than sixfold. The effect of the β and γ radiation is an additional small increase of ionization. Starting the experiment with careful observations of the normal rate of collapse of the leaves, the increase of this rate even by 50 per cent. is a variation safely beyond errors of observation. It will be presently seen that we possess additional assurance against error in the particular manner of growth of the increased rate of discharge due to the emanation of radium.

The reasons for the sensitiveness of the new method are obviously contained in the facts that we are here dealing with energy set free in considerable amount even by the disruption of one single atom, and that it is possible by the conditions described to utilize this energy in the creation of an ionic sluice or valve, whereby the relatively enormous pent-up energy in the gold-leaf is suffered to run down. This, it will be seen, is entirely different in theory from any method of measurement previously known to science. The feeble gravitational force on many trillions of atoms is required to incline the chemical balance. Here the kinetic energy of individual atoms derived from the internal stores of transforming atomic systems is called upon to operate the mechanism.

Since the investigation of the properties of the radium emanation by Rutherford and Soddy, and by the Curies, the estimation of radium has generally been effected by observations in which the emanation alone is introduced into the electroscope. The superiority of this procedure resides in the distinctive and convenient properties of this substance. It is the one gaseous element produced by radium in its sequence of transformations; and by suitable treatment of the material under investigation, if there is even the minutest amount of radium present, the equilibrium amount of emanation may be withdrawn from it and successfully introduced into the electroscope. The mere fact that it is the one gaseous body produced enables us to be sure that none of the other products of transformation is withdrawn along with it: these other elements remaining behind in the material under treatment if the experiment is rightly conducted. Knowing that the equilibrium amount of emanation is strictly proportional to the quantity of radium giving rise to it, we in turn are safe in assuming that the effects on the electroscope are proportional to the amount of parent radium.

To Boltwood[1] we owe a simple means of conferring on this method of measurement a strictly comparative character, so that we can calculate directly from the changed rate of collapse of the gold-leaf the amount of radium responsible for it. The emanation from a known amount of radium is introduced into the electroscope, and the instrument carefully observed. The rate of collapse read under this condition enables us to compute the radium-value, as we may call it, of the rate of collapse, so that thereafter, when an unknown amount of emanation is ionizing the air in the electroscope, we infer directly from the attendant rate of discharge the amount of radium giving rise to it. The

[1] Boltwood, *Am. Journ. Science*, 18th Nov., 1904.

instrument is, in fact, standardized or calibrated much as we calibrate a spring-balance by putting known weights upon it, and finding the weight-value of a distension through one centimetre. In such a purely comparative use of the electroscope, we are not required to investigate the electric capacity of the system; and so long as we preserve all the conditions of the standardizing experiment, subsequent readings are independent of the fixed details of construction.

From what has been said it will be gathered that the investigator in this line of research is much concerned with the properties of the emanation of radium. Except at temperatures approaching $-160°$ C., the emanation is a gaseous substance. It is a liquid below this temperature. Accepting 226 as the atomic weight of radium, that of emanation is 222. The volume of the emanation in equilibrium with one gram of radium is 0·585 cubic millimetres.[1] Its rate of transformation is such, that in 3·75 days it is diminished to one-half the original amount: it may therefore be stored for some hours (as when in the electroscope) without serious diminution. The short transformation periods of the three following products of change are noteworthy. Radium A, B, and C are all perishable in periods measured by minutes: consequently their equilibrium values are rapidly attained. Thus in a few minutes as much of radium A is generated by the emanation as can exist along with it; radium B rapidly increasing at the expense of A; and while B is growing it is also decaying into radium C. The

[1] Rutherford and Geiger, *Proc. R. S.*, 81, A, 172.

next product is a rayless substance of great longevity. We are therefore not concerned with it. The result of all this is that the rate of discharge of the electroscope, if observed a few minutes after the emanation is introduced, and again after about three hours, will be found to have increased some 30 to 40 per cent., due to the increase in quantity of radium B and radium C. After this lapse of time there is no further increase of the leak, all the successive products of change being in equilibrium with one another. This behaviour is characteristic of the emanation of radium. The emanations of thorium and actinium, possessing transformation periods measured in seconds, would either not be brought into the electroscope unless special precautions were adopted, but in the process of transfer would assume the solid form, or, if introduced, they would not show the characteristic rate of increase of ionization.

The emanation of radium is a remarkably stable body. It has been subjected to the most extreme chemical and physical conditions by Rutherford, Ramsay, and others. Sparking for hours with oxygen over alkali; heating to a red heat in contact with various substances; etc., had no measurable effect on its radioactive properties. Passed through tightly compressed cotton-wool it issued unchanged in amount. It is, in fact, purely gaseous in its properties, and diffuses like any other gas.

The first products of change—as well as the subsequent ones—are solids. These are deposited from the emanation upon any surface exposed to the gas.

The first three derivatives of the emanation give rise to what is called "the excited activity" or "active deposit." There is evidence that in the mixture of air and emanation in the electroscope, the particles of radium A, B, and C exist suspended for some time before they are deposited on solid objects; the process of deposition being hastened by the establishment of an electric field. Once deposited, a certain number of the rays emitted in course of further changes must penetrate the solids on which the particles rest, without effecting any ionization of the gas. The behaviour in the gas of the transforming particles appears to be a matter of some obscurity.[1] It is possible to learn from the electroscope the result of an experiment by observation of its leakage shortly after introducing the emanation, and without waiting for the full growth of the active deposit; but in this case we must calibrate for the same stage of growth; that is, we may read the rate of collapse, say, 15 minutes after introducing the emanation, and, if we have read the leak after the same interval when the standardizing quantity of emanation was introduced, we may by comparison of the readings determine the radium. The sensitiveness of the method is, however, reduced by this procedure, and the assurance derived from watching the phenomena of development of the deposit is to some extent lost.

The short-lived nature of the active deposit secures the instrument from the accumulation of radioactive materials, which otherwise must soon render it useless. If the air and the minute amount of emanation diffused

[1] Rutherford, *Radioactivity*, pp. 319, 395.

therein are removed after an experiment, the excited activity remaining soon dies out. The normal properties of the instrument are thus not sensibly affected by successive experiments.

The mode of extraction of the emanation corresponding to the radium in any substance, and the nature of the procedure in measuring it, will now be described in outline.

The rock (supposing the substance to be of this nature) is brought into solution. In general it is best to divide the elements of the rock into two solutions—an alkaline and an acid solution. They are stored in closed flasks for about three weeks, at the end of which time the emanation has accumulated till it is almost in equilibrium with the amount of radium present. Each flask is then placed in connexion with a glass receiver, which has first been exhausted of air—*R* in the figure, p. 26. A condenser, through which cold water can be circulated, intervenes between the receiver and the flask—*C* in the figure. When the connexion with the receiver is made, by relaxing the pinched rubber tube at *B*, the pressure falls in the flask and the solution is now boiled, the steam being condensed in the condenser and draining back into the flask. The ebullition is continued for about 30 minutes, during which time the air and emanation are boiled out of the solution, and collect above the liquid in the flask, in the condenser and in the receiver. When the burner is turned off, cold distilled water is run into the flask by the tubulure *E*, and the whole of the gaseous contents displaced up into the receiver, the water being allowed to rise into its lower tubulure. The receiver is now closed by pinching it off at *B*, and its contents are transferred to the electroscope. To effect this the electroscope is first exhausted of air, and then connected to the receiver, which is so arranged that

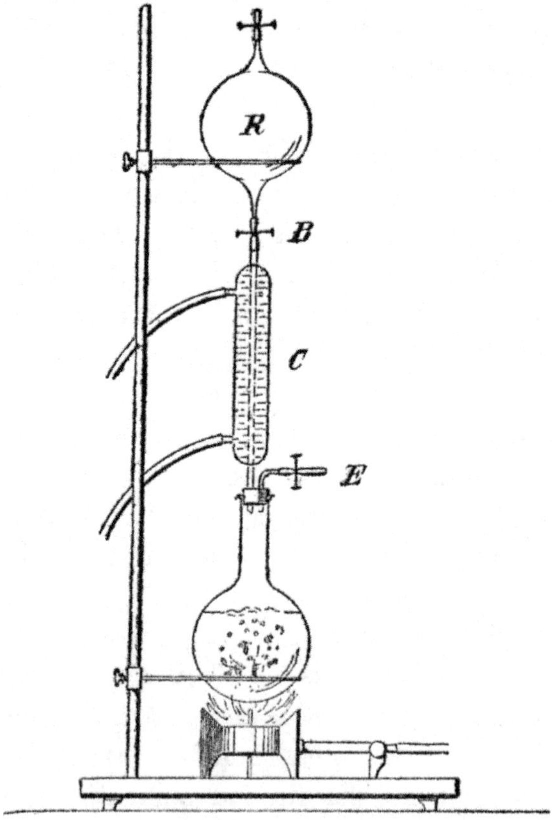

Fig. 1.

water flows into one of its tubulures while the electroscope draws the contained gas out by the other. Thus the whole of the boiled-off gas is brought into the electroscope without loss.

There are here three important conditions attained: ebullition *in vacuo*; collection of the emanation in a dry vessel; and complete transfer to the electroscope.

The electroscope is shown in fig. 2. It is a glass flask with a connexion, *a*, for leading gas into and out of it through the drying tube, *t*, which contains phosphoric anhydride. The gold-leaf system is very simple. A short strip of thin brass, about 2½ cms. long, and about 3 mms. wide, is sealed into the end

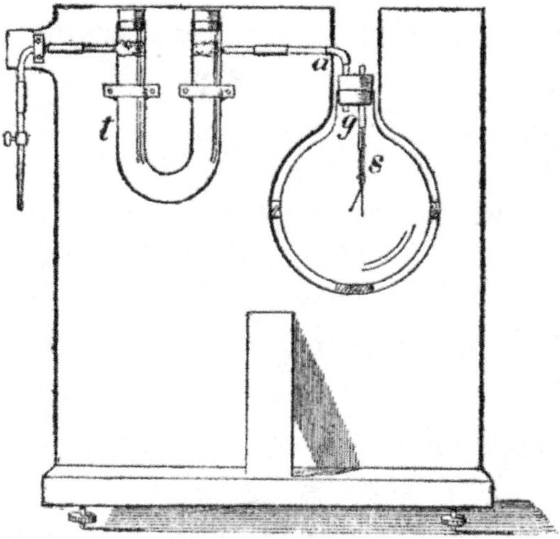

Fig. 2.

of a sulphur tube, *s*, of fine bore. This sulphur tube has been cast in the glass tube, *g* (which supports it in the stopper of the flask), and is then pushed forward in the glass tube, into which it is finally cemented air-tight. A narrow gold leaf is attached to the brass strip as shown. The system is electrified by inserting a platinum wire downwards through the bore of the sulphur tube till it makes contact with the end of the brass slip, the wire being then electrified by contact made above and withdrawn by means of an attached sulphur handle. The leaf should be made divergent by about 30 degrees. It is kept under observation by a reading microscope, having a divided scale in the eyepiece. The interior walls of the electroscope are connected to earth by two strips of tinfoil which enter beside the stopper.

The normal rate of discharge of the electroscope is first found and entered as so many scale-divisions (of the eyepiece) per hour. This rate remains practically constant if the instrument is pumped out after each experiment, and before making one. When the radioactive air which has been boiled off the solution is introduced, the rate of collapse of the gold leaf at once begins to increase. The new rate is noted. Again, in about three hours, in which time the active deposit is almost in equilibrium with the emanation, the rate is read. The increase of rate will be about 30 or 40 per cent. above the first reading.

The increased rate of discharge above the normal is converted to its radium value very simply on the knowledge of the constant of the instrument, which has already been determined as described. Suppose the constant tells us that an increased rate of collapse above the normal, of one scale-division per hour, corresponds to $0{\cdot}7 \times 10^{-12}$ grams of radium, and that the observed increase of rate in the experiment has been 40 scale-divisions per hour; then the radium in the flask must have been $40 \times 0{\cdot}7 \times 10^{-12} = 28 \times 10^{-12}$ grams. The alkaline solution is dealt with in the same manner, and the radium found added to that determined for the acid flask. The total radium is due to the rock which has been used, and to any radium which might be present in the chemicals used in preparing the solutions. The last allowance has been found by previous experiments on the chemicals, and after it is deducted the remaining radium is calculated as so much per gram of the rock.

Several conditions have to be attended to if reliable measurements are to be made. Some of these, as involved in the apparatus, have been referred to above. The most important remaining ones are:—(1) absolute avoidance of any possibility of contamination by radioactive preparations; (2) the use of the least amount of chemicals possible in preparing the solutions; and (3) complete absence of any precipitate from the solutions; which last condition involves (among other things) the use of hydrochloric acid which has been redistilled over sodium chloride in the laboratory, and contains no trace of sulphuric acid.

The principles involved in the method of measuring small quantities of radium will now be understood. Perhaps full confidence in its results will only be felt by those who know how repeatedly verified under various conditions, and under various forms, the laws of radioactive transformation have been. On these the whole method rests. And again, from the practical aspect the sensitiveness and constancy of the electroscope can only be appreciated by practice in its use. In a later chapter the essential practical details are given, mainly in the hope that the work of research in this branch of science may be extended, but partly in the belief that it is only in the laboratory that science can be rightly studied, and that those who cannot work in laboratories may find clearer views and fuller confidence by more closely following the details of an experiment upon which so much is founded.

In this chapter a few outstanding questions have yet to be dealt with.

The idea that the rapid heat-production and continued radiation of radioactive bodies may in some manner be enhanced by the concentrated form in which the radioactive salts are dealt with in the laboratory, and that consequently in nature, where there is enormous dilution with inert matter, similar actions might be much weaker, has, undoubtedly, a certain plausibility. For we would be inclined to expect that a certain amount of auto-excitation of radioactivity might arise from the rays produced. There is, however, a large body of experimental evidence against this view. Rutherford has directly sought for a change in the radioactivity of a radium salt when diluted as much as 1000 times, and could detect no change. If there was any such effect as auto-excitation, it should be apparent here. Eve has investigated some alleged positive results in this direction, and not only failed to confirm them, but found that in every case the activity was directly proportional to the amount of radium present within the limits of experimental error. Again, the radioactive constant or rate of transformation has been investigated for many substances over wide ranges of concentration, and never known to vary. In the case of radium emanation, the rate of decay has been observed when the concentration in air was varied from one to a millionfold, but no difference was discoverable.

But, finally, as regards the geological applications of radioactivity, if we reflect on the nature of our observations when we withdraw emanation from rock materials which have been brought into solution, we perceive that we are not really concerned with the question of concentration. The solution represents even a more dilute radioactive medium than does the original rock. The emanation we withdraw must have been generated in it by the radium present, for in the process of preparing the solution we de-emanate it completely. But if formed in the solution, then must it be formed in the rock; and hence we have proof of the instability of the radium and its progenitors. Again, as the emanation does not accumulate indefinitely in the solution, it must be transforming; and so, practically, the whole series of changes is established, and with it the heat-production which is but the outcome of the absorption of the several rays in the medium.

Following this same line of reasoning, the verification of the law of the constant proportion of radium to uranium in substances where the concentration was enormously different, is again a proof that concentration has nothing to do with the production and continuance of radioactivity. Now, in Boltwood's experiments this fixed ratio, which, as we have seen, depends entirely on the constants of change of the two substances, has been verified when the quantity of radium present in one gram of the ore must have been about 3×10^{-7} grams. And in an experiment by E. Werner and the author, this constant ratio was found to hold in a rock in which the radium was present only to the amount of 5×10^{-11} grams per gram of rock. The concentration of the radioactive elements in the first experiment is 10,000 times that in the second, yet the change rate has been unaffected. There can be no question about the conclusion: the geological applications of radiothermal actions are not restricted by any limitations arising from concentration.

But they might be restricted in yet another manner. When we come to apply the facts we have been studying to the large questions of terrestrial physics, it is necessary to consider if the observations of the laboratory carried out at normal air-temperatures and pressures are applicable to the physical conditions which we believe on good grounds must prevail in the interior of the earth.

From time to time the independence of radioactive phenomena of physical and chemical conditions has been brought out in experiments. Those already cited are in point. More lately it would appear as if radium C is in a small degree an exception, although, as will be seen, there is conflict of testimony. Bronson,[1] using radium and its products, could detect no change in the rates of transformation even after prolonged exposures

[1] Bronson, *Proc. R. S.*, 78 A, 494.

to temperatures up to 1600° C. and so low as – 180° C. Observations by Makower,[1] while agreeing that there is no change in the emanation, appear to show a small change in radium C, as measured by the β and γ rays, when exposed to temperatures of about 1100° C.

Schuster,[2] as well as Eve and Adams,[3] using direct mechanical pressures, found that the highest attainable pressures were without influence on the activity of radium. Rutherford and Petavel[4] employed the intense temperatures and pressures arising from an explosion of cordite—about 2500° C. and 1200 atmospheres—to find if any change in the radiation of the active deposit could be observed; the emanation of radium being enclosed along with the explosive in a steel explosion chamber, the γ radiation passing through the walls to the electroscope. It was found that while the radiation was not affected during the explosion, a decrease of temporary nature to the extent of about 9 per cent. in the γ radiation was indicated after the explosion.

More recently still H. W. Schmidt,[5] in experiments undertaken to decide between the previous contradictory results as to change in radium C. at high temperatures, using a special preparation of this derivative of radium, found that at 1500° C. this substance behaves exactly as at ordinary temperatures. At pressures of 2000 atmo-

[1] Makower, *Proc. R. S.*, 77 A, 241 and 79 A, 158.

[2] Schuster, *Nature*, July, 1907, p. 269.

[3] Eve and Adams, *ibid.*

[4] Rutherford and Petavel, *Rep. Brit. Assoc.*, 1907, p. 456.

[5] Schmidt, *Physik. Zeit.*, Feb. 15, 1908. See also *Ion*, Nov., 1908, p. 70.

spheres no change in the intensity of the radiations of radium preparations was perceived.

As the experimental evidence stands, the constancy of the radioactive processes seems proved for radium and its emanation over wide limits of temperature and pressure such as would obtain deep within the surface-materials of the earth. The question of the constancy of the radioactive change of the parent substance, uranium, has not been treated in the laboratory; but the argument from analogy is all against any departure from the known results. There is, indeed, nothing in the evidence at our disposal to prohibit us from supposing that radioactivity may exist unchanged in its laws even to the centre of the earth. True, the experimental conditions of pressure would hardly apply below 50 miles downwards—a very small fraction of the earth's radius—and the temperature conditions reached in the laboratory might or might not be such as prevailed at the central parts of the globe; but it seems most probable that the laws of radioactive change are based on properties of the atom inherent in its intimate structure, and on energy conditions so self-contained that the internal relations of the atom with the ether alone affect its readjustments and its life-history. Modern theory certainly tends to regard radioactive phenomena as due to events progressing within the atom, and, as such, distinct from chemical actions, which involve external effects upon surrounding atoms, and hence the physical conditions under which the atoms are brought together.[1]

[1] See Lodge, *Nature*, June 11, 1903, p. 129; and Thomson, *Phil. Mag.*, Dec., 1903, and March, 1904.

CHAPTER III.

THE RADIUM IN THE EARTH'S SURFACE-MATERIALS.

EARLY OBSERVATIONS—THE RESEARCHES OF STRUTT—THE RADIO-ACTIVE SUBDIVISION OF ROCKS—THE RADIOACTIVITY OF IGNEOUS ROCKS—RADIOACTIVITY OF THE OCEAN—IONIZATION OVER THE SEA—RADIOACTIVITY OF DEEP-SEA DEPOSITS—VARIATION WITH AMOUNT OF CALCAREOUS MATTER—URANIUM NECESSARILY IN THE OCEAN—THE RADIUM IN THE NILE AND IN THE RIO DE LA PLATA—URANIUM IN THE RED CLAY—ESTIMATE OF THE MASS OF SUB-OCEANIC SEDIMENTS AND OF THE QUANTITY OF RADIUM THEREIN—THE DETRITAL ROCKS—SEDIMENTS DERIVED FROM SOLUTION—NO SIGN OF DIMINISHING RADIOACTIVITY WITH THE DEPTH—THORIUM AND HELIUM IN THE ROCKS—PLEOCHROIC HALOES.

THE gaseous nature of the emanation derived from radium was instrumental in first indicating the very widespread existence of radioactive substances in the surface-materials of the earth. In 1902 Sir J. J. Thomson found that air bubbled through the Cambridge tap-water gained in conductivity. In 1903 Elster and Geitel observed that air drawn through the soil contained a radioactive emanation; and they separated chemically a radioactive substance from clays and soils. Consistently with these observations it was found that the air in caverns was exceptionally radioactive. The radioactive emanation was also found to be generally diffused through the atmosphere, depositing an active substance on negatively charged surfaces. These and a host of observations by various investigators all over the world pointed to the conclusion that radioactive matter was a widely diffused constituent of the surface-materials of the earth.

Speculation was awakened by such a far-reaching conclusion: and Rutherford and Soddy[1] and W. E. Wilson independently suggested even a cosmic application—the possibility that radiothermal actions may contribute to solar heat. Sir George Darwin[3] pointed out the influence which such a view must have on speculations as to the age of the sun. The present writer[4] drew attention to its probable importance as a source of terrestrial heat, whether only abundant in the crust or deeper down; and the effect of this view on conclusions as to the earth's age which had been founded on the existing thermal state of the crust. Finally, Rutherford calculated that a quantity so minute as $4{\cdot}6 \times 10^{-14}$ grams per gram of the earth's materials would account for all the heat which, on the evidence of the surface-gradients of temperature in the crust, we know to be escaping at the surface.

No even approximate estimate of the concentration of radium in the rocks and soils was possible, however, till the appearance of the Hon. R. J. Strutt's researches in the year 1906.[5] For the first time definite quantitative measurements on rocks and soils gathered from widely separated localities were here forthcoming. This work also first established the fact that the quantities at the surface of the earth are much in

[1] Rutherford and Soddy, *Phil. Mag.*, May, 1903.
[2] Wilson, *Nature*, July 9, 1903.
[3] Sir G. Darwin, *Nature*, September 24, 1903.
[4] Joly, *Nature*, October 1, 1903.
[5] Strutt, *Proc. R. S.*, 77 A, 472 and 7 A, 150.

excess of Rutherford's estimate quoted above. Consequently a purely surface-distribution was assumed by Strutt, extending to a depth of about 45 miles. This estimate would not, however, indicate the true bearing of the experiments according to the views of the investigator; for, owing to an erroneous idea prevailing at the time as to the constant proportion of radium to uranium, which constant enters into the calibration of the electroscope, the results obtained by Strutt require to be reduced by nearly half their value. Hence, on the theory put forward by Strutt, a layer of about twice this depth must be required to account for the earth's heat.

The measurements made by Strutt, as corrected for the error just referred to by Eve,[1] are as follows:—

IGNEOUS ROCKS.

		Radium in grams per gram.
Granite,	Rhodesia,	$4{\cdot}78 \times 10^{-12}$
,,	Cornwall,	4·67 ,,
Zircon syenite,	Norway,	4·65 ,,
Granite,	Cornwall,	4·21 ,,
,,	Cape of Good Hope,	3·57 ,,
,,	Cornwall,	3·45 ,,
,,	Westmoreland,	3·31 ,,
Syenite,	Norway,	2·44 ,,
Granite,	Devon,	1·84 ,,
Blue ground,	Kimberley,	1·68 ,,
Leucite basanite,	Vesuvius,	1·66 ,,
Hornblende granite,	Egypt,	1·22 ,,
Pitchstone,	Isle of Eigg,	1·03 ,,
Hornblende diorite,	Heidelberg,	0·99 ,,
Augite syenite,	Norway,	0·93 ,,

[1] Eve, *Phil. Mag.*, August, 1907, p. 231.

IGNEOUS ROCKS—*continued*.

		Radium in grams per gram.
Peridotite,	Isle of Rum,	$0 \cdot 68 \times 10^{-12}$
Olivine euchrite,	,,	0·64 ,,
Olivine basalt,	Skye,	0·66 ,,
Basalt,	Victoria Falls,	0·63 ,,
Hornblende granite,	Leicestershire,	0·62 ,,
Dolerite,	Isle of Canna,	0·62 ,,
Greenstone,	Cornwall,	0·57 ,,
Basalt,	Antrim,	0·52 ,,
Serpentine,	Cornwall,	0·50 ,,
Granite,	Isle of Rum,	0·36 ,,
Olivine rock,	,,	0·33 ,,
Dunite,	Loch Scaivig,	0·33 ,,
Basalt,	Greenland,	0·30 ,,

SEDIMENTARY ROCKS.

Oolite,	Bath,	2·92 ,,
,,	St. Alban's Head,	2·02 ,,
Marble,	East Lothian,	1·93 ,,
Kimmeridge clay,	Ely,	1·88 ,,
Oil-bearing sandstone,	Galicia,	1·52 ,,
Roofing slate,	Wales ?	1·28 ,,
Gritty slate,	Cornwall,	1·25 ,,
Gault clay,	Cambridge,	1·01 ,,
Clay,	Essex,	0·86 ,,
Red sandstone,	East Lothian,	0·84 ,,
Fine gravel,	Essex,	0·71 ,,
Red chalk,	Hunstanton,	0·53 ,,
Flint,	Essex,	0·53 ,,
White marble,	India,	0·27 ,,
Marble,	East Lothian,	0·26 ,,
Chalk, bottom of pit,	Cambridge,	0·39 ,,
,, top of same pit,	,,	0·12 ,,

Eve finds the mean of these results to stand as follows :—

Igneous rocks, ...	$1 \cdot 7 \times 10^{-12}$
Sedimentary rocks, ...	$1 \cdot 1$,,

In 1907 Eve[1] made some determinations on Canadian rocks. The results arrived at were :—

	Radium in grams per gram.
Trenton limestone, crystalline, Ordovician,	$0 \cdot 92 \times 10^{-12}$
,, ,, weathered, ,,	$0 \cdot 91$,,
Essexite. Devonian ?	$0 \cdot 26$,,
Tinguaite, ,, ...	$4 \cdot 30$,,
,, different locality. Devonian ?	$3 \cdot 00$,,
Nepheline syenite. Devonian ? . . .	$1 \cdot 10$,,
Boulder clay, Quaternary, ...	$0 \cdot 80$,,
Leda clay, ,, ...	$0 \cdot 88$,,
Saxicava sand, ,, ...	$0 \cdot 16$,,

These rocks are from the neighbourhood of Montreal.

The view that *all* the radium is collected near the surface in such concentration as the rocks indicate is beset with difficulties to which we must refer in another place. But that there must be a considerable degree of surface-concentration admits of no doubt. This radioactive surface-layer must extend to a depth of some miles. This at first sight seems inexplicable. For the parent substance uranium is the heaviest element known, and its concentration near the centre of the earth is more what we would have expected. The consideration and possible explanation of the observed facts we must postpone. It is sufficient for us at present

[1] Eve, *loc. cit.*

to recognize that our surface investigations cannot apply to the whole bulk of the earth, and that in view of the importance attached to the downward distribution of radium, measurements effected on igneous materials, probably coming from considerable depths, and on such materials as have been encountered in the deepest borings and tunnels, possess particular interest. The problem, too, of the general surface-richness of the earth is closely associated with the radioactive state of the ocean and its sediments.

From the radioactive point of view the surface-rocks may be divided into :—

(*a*) Igneous; possessing a store of uranium presumably coeval with the association of elements which led to the formation of the originating magma.

(*b*) Detrital sedimentary; the more or less disintegrated and granular remains of rocks of group (*a*). The uranium in these rocks is mainly residual in the case of the coarser varieties, as sandstones, breccias, &c.; but in the finer varieties, as slates and fine shales, the uranium was probably derived in part by precipitation from the waters in which the sediment was deposited.

(*c*) Precipitated sedimentary; in which the whole or nearly the whole of the materials has been derived from dissolved substances by organic or purely chemical agencies. In these the uranium is in most cases derived entirely from the waters in which the rocks were formed.

The results obtained by the author on the radioactivity of rocks are in general agreement, and in many cases in close agreement, with those obtained by Strutt and Eve. It is probable that what difference is exhibited may arise from the natural advance in our knowledge of the best conditions of extracting the emanation. Thus it has been experimentally found that any trace of precipitation may conceal the emanation, retaining it in the solution even after prolonged boiling. Again, it seems certain that ebullition *in vacuo* and collection in a dry vessel is the more effective procedure in carrying out the extraction. The concentration of large amounts of dissolved substances in the solutions is also a source of diminished results. It is important, too, to operate on fresh, unweathered specimens of rock. A summation of these influences probably results in the higher general means recorded further on.

The applications of the surface-radioactivity of the earth to geological problems, which occupy most of the ensuing chapters, are based on values somewhat below the means of all published results.

The basic rocks of the Deccan, which are of enormous extent, have yielded radium at every point at which they have been examined. A series of tunnel and surface specimens, kindly sent by the Director-General of the Indian Geological Survey—Sir Henry Holland

—enabled estimates to be made at widely separated points. The specimens examined up to the present are the following:—

Bedded basalt, Ghats near Thuria, ...	2·7
,, ,, Pavagad Hill, Panch Mahals, .	8·2
Basalt dyke, near Palasi, Dhar State (2), . .	4·9
Bedded basalt, tunnel near Palasdari, . . .	3·8
,, ,, ,, between Keloli and Vani, .	2·4
,, ,, ,, between Kasara and Igatpuri,	3·6
Mean, . . .	4·3

The convention of omitting the qualifying expression 10^{-12} will be continued, as above, throughout the rest of this book unless occasion arises to specially refer to this factor. It will be, therefore, always understood that the radioactivity expressed by an unqualified number is the quantity of radium in billionths of a gram per gram of material under discussion. A number given in brackets after the designation of a substance refers to the number of tests, generally on independent samples of the substance: the radioactivity in such cases being the mean of the results.

The basalt of the Giant's Causeway, which is part of the great lava flow of the Inner Hebrides, has been treated in five experiments, all, however, on adjoining fragments of the rock. The results agree very closely. In all there are fourteen results on basic rocks:—

Deccan basalts and traps (6), ...	4·3
Giant's Causeway basalt (5), ...	5·8
Basalt, Mellenifjord, Greenland, ...	5·0
"Whinstone," Balfour Bore, Fifeshire (from a depth of 3000 feet), ...	3·8
Mean, ...	4·9

On granites there are six results :—

Leinster granite, Ballyknocken (2), ...	5·4
Mourne granite, Newry, Co. Down, ...	3·7
Aberdeen granite (3) (from depths of 80 to 270 ft. in Kemnary Quarry), ...	3·2
Mean, . . .	4·1
Syenite, Plauen, ...	6·8

And the following from the very ancient rocks of Sutherlandshire, Scotland :—

Lewisian mica schist, ...	5·3
,, gneiss, ...	3·8
,, hornblende schist, ...	8·0
Mean, . .	5·7

Twelve specimens of recent volcanic rocks :—

Lava, Crater, Kilauea, 1875, ...	6·2
Scoria, Stromboli, ...	6·0
Augite andesite, Val di Lachi, Lipari, . . .	2·4
Trachyte, Monte Olibano, Campi Phlegraei (2), .	10·6
Obsidian, Ascension Island, ...	2·5
Lava, Rocci Rossi, Lipari (2), ...	8·0
Bomb, Vulcano (2), ...	7·6
Olivine trachyte, Ischia (2), ...	9·4
Pumice, Chimborazo, near Quito, ...	4·8
Lava, Vesuvius, 1855 (2), ...	19·2
Bomb, Martinique, 1902 (weathered), . . .	2·0
Pumice, Krakatoa, ...	6·9
Mean, . . .	7·1

To these must be added a large number of rocks which enter into the composition of the Simplon massif, and which are believed to be of igneous origin; and others again which enter into the St. Gothard and the Finsteraarhorn massifs. These are in every case tunnel specimens, some of those from the Simplon coming from depths beneath the surface of 2135 metres (7005 feet).

Reserving particulars till we come to consider specially the results obtained in the case of these two great tunnels, it is sufficient to say here that in the case of the Simplon some difficulty was experienced from what may be called sporadic radium. Possibly the cause of this has to do with the facts that not only are many of these rocks rather above the average in radioactivity, but they have been exposed to extreme metamorphic actions which may have led to a segregation of the uranium. Some, bearing evident uranium minerals, were met with, so that a low and a high mean, the latter embracing a few very exceptional results, appear to arise out of the experiments. Further experiments are in progress to determine, if possible, the most probable average. Such extreme variations were not noticed in the St. Gothard rocks. The Finsteraarhorn granite is included in the mean of the St. Gothard igneous rocks.

Simplon (32);	schists and gneisses, mean, low, .	7·6
,, (32);	,, ,, ,, high, .	9·1
St. Gothard (32);	granites, gneisses, and schists, mean,	5·1

The mean of all the igneous rocks examined by the author (92 distinct specimens), taking the low mean of the Simplon rocks, is 6·2. This may be somewhat unduly raised owing to the seemingly rather abnormal results obtained with the Simplon materials. Setting aside for the present the entire Simplon series of igneous rocks, the mean of 60 rocks is 5·5. These 60 results, taken along with those of Strutt and Eve on similar materials, give a general mean of 4·2 obtained on 92 specimens of various igneous rocks.

It would be impossible to predict with any certainty the relative radioactive states of the sediments and the parent igneous rocks. So far as the purely detrital rocks are concerned, the question would turn, of course, on how far the uranium is retained in the debris of the parent rock, and on the weathering qualities of certain minerals which appear to be very often specially radioactive; for instance, zircon. There is little doubt that high results occasionally met in rocks of the detrital class are due to the preservation of this mineral. Evidently we could not safely predict the result where such uncertain factors enter; but, reasoning from the behaviour of other heavy metallic substances, the probability would incline in favour of the uranium being in considerable part removed in the general disintegration of the rock. Thus we should expect a lowered radioactivity in the case of such rocks. The precipitated sediments present a very different problem, and one which experiment alone could solve. Here the radioactivity must depend on the abstraction of uranium along with the rock-materials in the process of deposition, and clearly the radioactivity of the ocean is concerned in the genesis of such materials. The uranium in the ocean and its sediments is indeed a factor of such considerable importance in the consideration of the whole question of radioactive denudation, seeing that the whole or the greater part of it must have come from the land, that we will state what is known on the subject before passing on to the sediments.

The waters of the ocean cover five-sevenths of the earth's surface to a mean depth of about 3·8 kilometres, and represent, of course, far the most abundant of the surface-materials open to our investigations. A considerable number of results have accordingly been obtained in order to arrive at some conclusion as to the probable mean radioactivity. Quantities of the water from 1500 to 3000 c.cs., in most cases concentrated to about half bulk by evaporation in a laboratory into which radioactive preparations had never come, were used for testing. The results obtained are for the greater number published with fuller particulars elsewhere.[1]

	Radium per c.c.
Coast round Ireland (5), ...	0·034
Atlantic Ocean, Madeira to Bay of Biscay (5),	0·017
Atlantic, New York to Ireland, Arctic current,	0·016
,, ,, ,, ,, Gulf Stream, .	0·014
,, ,, ,, ,, Mid Atlantic, .	0·011
,, ,, ,, ,, 260 miles W. of Ireland, .	0·008
,, lat. 0° 0′, long. 31° 26′ W., ...	0·038
Atlantic, South; lat. 16° 54′ S., long. 37° 49′ W.,	0·007
Mediterranean (2), ...	0·008
Black Sea, ...	0·007
Arabian Sea, lat. 10° 40′ N., long. 58° 0′ E., .	0·027
Indian Ocean to Mediterranean—	
Sandheads to Madras,	0·004
,, ,, Off Colombo, . . .	0·007
,, ,, Minicory to Sokstoa,	0·004
,, ,, Red Sea, off Jiddah,	0·009

[1] Joly, *Phil. Mag.*, March, 1908, p. 385.

The mean of the twenty-four determinations, many of which are individually the results of several experiments, is $0{\cdot}017 \times 10^{-12}$ grams per cubic centimetre.

These results are obtained by modes of treatment varied in such a manner as would best guard against any systematic experimental error. Sea-waters are only safely dealt with on addition of pure redistilled hydrochloric acid, which lessens greatly the risk of a precipitate forming in the process of concentration. The variability of the results is considerable. To what is this due? It would seem unlikely that retention of the emanation can account for the lower results, for most of them were confirmed by a second boiling, and in some cases on three or four repetitions of the experiment. The second boiling generally gives a somewhat increased yield of emanation, but there is seldom any subsequent rise. Again, the matter was investigated by adding to a sea-water a certain known small quantity of radium, and seeking if the emanation corresponding to it could be completely withdrawn. The result was quite satisfactory. This experiment does not indeed prove conclusively that a residuum of emanation might not remain over in all sea-waters after ebullition; but if so, we would expect this residuum to be fairly constant in amount, and the differences from experiment to experiment have still to be explained. Nothing is more astonishing than the fact that these small amounts of emanation can, in any case, be withdrawn. The volume of the emanation per gram of radium is but 0·585 cub. mms.[1] Thus there are

[1] Rutherford and Geiger, *Proc. R. S.*, 81 A., 173.

but 9 billionths of a cubic millimetre in a litre of average sea-water!

It is, perhaps, premature to say much as to the possible cause of the variations. The most plausible explanation seems to be the removal of radioactive matter from the surface-waters by precipitation under the action of decaying organic life and sulphur bacteria; so that according to the local organic conditions—and it is known that these vary considerably under the influences of surface-currents, etc.—a large variation of the radium near the surface might arise. The investigation of water from the deeper parts of the ocean has yet to be carried out, and will be of interest.

There is evidently a large quantity of radium in the ocean. Taking the mass of the ocean as $1{\cdot}452 \times 10^{18}$ tonnes, there must be about 20×10^{9} grams, or about 20,000 tons of radium, in its waters.

It is of interest to note in connexion with the radioactivity of sea-water that Eve[1] has found that the ionization of the atmosphere over the central parts of the Atlantic is comparable with that over the land, even in the calmest weather, when it cannot be supposed that the emanation, which, doubtless, is responsible, could have been blown from the continents. This fact appears to be traceable to the radioactivity of the water, and the facility with which emanation may be supposed to escape from it. The bubbling of air through liquids containing a trace of radium results in loading the air with the emanation.

[1] Eve, *Phil. Mag.*, 6, xiii., p. 248.

Hence every bubble rising through the water, or mingled with it in rough weather, brings emanation to the surface. Again, it is certain that the emanation diffuses out at the surface; and as this is continually being renewed by currents and disturbances of every kind, the effect must be maintained continually at its maximum. It must result from these causes that the ocean, although not nearly so radioactive as the land, must give up its emanation more freely, and hence the observed ionization appears accounted for.

The condition of the deep-sea deposits fully supports the observations on the radioactivity of the ocean. From materials supplied to the author, mainly by Sir John Murray, seventeen various samples of oceanic deposits have been examined.[1]

Blue Mud. "Challenger." Off E. coast of N. America. Lat. 38° 34′ N., long. 72° 10′ W. 1240 fathoms. Radium 3·1.

Terrigenous Mud. From "mud volcano," coast of Arakan, Bay of Bengal. Surface. Radium 2·9.

Green Sand. Ss. "Albatross." Locality unknown. Radium 4·5.

Globigerina Ooze. Off W. coast of Ireland. 570 fathoms. Radium 6·6.

Globigerina Ooze. Central N. Atlantic. Lat. 48° 17′ N., long. 39° 49′ W. 2493 fathoms. Radium 7·0.

Globigerina Ooze. "Challenger." Middle of S. Atlantic. Lat. 21° 13′ S., long. 14° 2′ W. 1990 fathoms. Radium 6·7.

[1] Joly, *Phil. Mag.*, July, 1908, p. 190.

Globigerina Ooze. "Challenger." Pacific Ocean, W. of S. America. Lat. 38° 6′ S., long. 82° 2′ W. 1825 fathoms. Radium 7·4.

Globigerina Ooze. "Albatross." Central Pacific. Lat. 22° 11′ S., long. 133° 21′ W. 2042 fathoms. Radium 8·0.

Calcareous Mud. "Albatross." E. of Society Islands. Lat. 21° 4′ 6″ S., long. 133° 1′ 2″ W. 2225 fathoms. Radium 22·2.

Red Clay. "Albatross." N. Pacific, W. of Central America. Lat. 10° 38′ N., long. 105° 47′ 6 W. 1955 fathoms. Radium 13·0.

Red Clay. "Challenger." N. Atlantic, off coast of Africa. Lat. 24° 20′ N., long. 24° 28′ W. 2740 fathoms. Radium 15·4.

Red Clay. "Challenger." Central Pacific, near region of Radiolarian Ooze (as under). Lat. 13° 28′ S., long. 149° 13′ W. 2350 fathoms. Radium 52·6.

Radiolarian Ooze. "Challenger." Central Pacific. Lat. 3° 48′ S., long. 152° 56′ W. 2600 fathoms. Radium 22·8.

(From this sample the magnetic particles had been abstracted.)

Radiolarian Ooze. "Challenger." Central Pacific. Lat. 7° 25′ S., long. 152° 15′ W. 2750 fathoms. Radium 50·3.

Manganese Nodule. Locality same as that of last ooze. Radium 24·0.

Manganese Nodule. "Albatross." Pacific, off W. coast of S. America. Lat. 8° 29′·5 S., long. 85° 35′·6 W. 2370 fathoms. Radium 21·0.

The diatom oozes have not yet been examined.

The great quantity of radium spread over the ocean floor is well shown when we enter the above results against the extension of the several sediments as determined by Sir John Murray :—

	Radium.	Extension : millions of square miles.
Globigerina Ooze, . .	7·2 . .	49·5
Radiolarian Ooze, . .	36·7 . .	2·5
Red Clay, ...	27·0 . .	51·5

It is established on satisfactory evidence that the more slowly collecting deposits are those of non-calcareous character—the red clay and the radiolarian ooze; the red clay in some cases being almost devoid of calcareous matter, and collecting at an extremely slow rate. The evidence for this is the amount of unburied and even fossil remains which are brought up with such deposits. The foregoing results show that these deposits are richer in radium than the more rapidly growing calcareous sediments. This would follow if we supposed the rain of calcareous and siliceous tests acted as a dilutant. It is of interest to note in this connexion that the process of extraction of lime salts from the ocean by organic actions must, to a considerable extent, exclude the uranium in the waters, as indeed might be anticipated. The curious calculation of Bischof,[1] that an oyster requires the lime from some 27,000 to 76,000 times its weight of sea-water, would establish this fact; for if the uranium present in this bulk of water was taken into the shelly limestones, their radioactivity would be much greater than experiment shows.

[1] Bischof, *Elements of Chemical and Physical Geology*, vol. i., p. 180, English translation.

The following table shows that radioactivity and the percentage of calcareous matter in these deposits stand in an inverse ratio :—

		Calcium Carbonate per cent.	Radium.
Globigerina Ooze.	"Chall." 338,	92·24	6·7
" "	" 296,	64·34	7·4
Red Clay,	" 5,	12·00	15·4
"	" 276,	28·28	52·6
Radiolarian Ooze,	" 272,	10·19	22·8
" "	" 274,	3·89	50·3

The percentages of calcium carbonate are from the report of the "Challenger" Expedition. The red clay fourth from the top in the table might appear exceptional to the others, having too high a percentage of calcium carbonate for the radium content. It is probable that this is a case of recent change in the character of the deposit, the antiquity being shown by the number of sharks' teeth and manganese nodules brought up. Thus the specimen is a mixed one, so to speak, there being a lower part rich in calcareous matter, which acts to dilute the red clay more recently forming above. Several such cases of recent change are recorded by Murray and Renard.[1]

The view that the radium on the earth's surface might be of extra-terrestrial origin[2] suggested the desirability of examining the cosmic particles long ago identified by Sir John Murray as widely distributed over the ocean floor, and, as it were, concentrated in the deposits of very slow growth. If the theory

[1] *Report Scientific Results*, "*Challenger*" *Expedition*.

[2] Joly, *Nature*, January 24, 1907, p. 294.

has any reality, we should expect that such bodies would reveal exceptional richness in uranium, and, of course, in radium. A tube of magnetic particles, among which the cosmic particles would be included, received from Sir John Murray, was accordingly tested for radium. These magnetic particles had been withdrawn from material raised from a depth of 2300 fathoms in the North Pacific. The quantity available was only 1·1 gram. The result was a radium content, with difficulty read, of 0·6. Had there been present even one rich uranium-bearing particle, so small a result would not have been obtained. The theory referred to therefore lacks any support which this experiment might afford.

We have seen in an earlier chapter that it is to the genetic connexion between uranium and radium that we look for the maintenance and renewal of the latter. It is thus of importance, in relation to oceanic radium, to establish, where practicable, the existence of the parent substance, and thus obtain assurance that the radium is not exotic in character, and the radioactivity a passing property. In the majority of rock-investigations this is impossible, owing to the minuteness of the quantities of uranium involved, and the lack of character which the uranium spectrum exhibits. But in these cases the conditions assure us that, unless there be some other parent to radium, the uranium must be there; for how else has the radioactivity been maintained through past ages? The oceanic conditions lead us to a similar conclusion, so far as the waters of the ocean are concerned. In the great volume of the ocean we have estimated not less than 20,000 tons of radium. This cannot have been supplied in a period so brief as to be included in historical times. Yet, unless the uranium is present, the whole amount of radium must be renewed in a few thousand years by the rivers. In order to accomplish this, however, the rivers must be much richer in radium than the ocean. We can acquire an idea of what their radium content must be. The waters of the ocean possess a total mass of $1{\cdot}4 \times 10^{18}$ tonnes. The amount of water annually discharged by the rivers of the world has been estimated by Murray at 27,191 cubic kilometres, or about 27×10^{12} tonnes. Now $\frac{1}{2540}$ part of the radium is transformed each year, so that if the rivers are effective in keeping up the supply, they must in one year send into the ocean an amount of radium equal to $\frac{1}{2540}$ of what is present in its waters. But the quantities of ocean-water and river-discharge per annum are as 50,000 to 1, or, in other words, the rivers must take 50,000 years to replenish the ocean. They should, therefore, be richer in radium in the ratio of 50,000 to 2540, or as, say, 20 to 1. There is no evidence of river-water possessing such richness. An examination of Nile water afforded a radioactivity of but 0·0042, which is about one-fourth the oceanic, and water from the Rio de la Plata showed 0·0052.

Again, we are obviously in the difficulty of accounting for the uranium corresponding to the radioactivity of the ocean if not conveyed to the sea. For in this case it must be in the detrital sediments, and these should show the concentrated radioactivity of ages of denudation.

This they do not, but in point of fact are less radioactive than the igneous rocks—a state which is in harmony with one conclusion only: that the uranium released from the rocks by denudation is carried into the ocean.

While this reasoning seems unassailable as regards the ocean, we cannot apply it rigorously to the sediments accumulating beneath. It might be that a certain precipitation of radium would give to the surface parts of these deposits a temporary high radioactivity.

Against this view we have the fact that such ancient deep-sea deposits as are open to our examination show considerable radioactivity. The chalk and greensand are instances. The former must have been a relatively fast-formed deposit composed almost entirely of substances extracted from solution; its radioactivity compares well with that of Globigerina Ooze. A specimen of green sandstone showed a somewhat higher radioactivity than a similar sediment recently dredged. Plainly in these cases the uranium must have been precipitated along with the other materials.

As further evidence on this point, in the case of a red clay showing the high radioactivity of 54·4, Professor Emil Werner has recently estimated the uranium, using a colorimetric method. In 8·42 grams of the clay submitted to Werner, the quantity of uranium should be 1·3 milligrams. The estimate afforded from 0·6 to 0·7 milligrams. But the separation of the uranium is probably incomplete.[1]

If we take it that the richness in radium of the

[1] Joly, *Phil. Mag.*, July, 1908, p. 196.

oceanic sediments is on the average no greater than that which we observe in the Globigerina Oozes—and this is probably a low estimate—the quantity of radium involved is enormous. We can calculate with considerable security the aggregate mass of such oceanic deposits as are derived from dissolved materials; *i.e.* the calcareous oozes and muds, the diatomaceous oozes, the radiolarian oozes, and much of the mass of the red clays.

The elements of this seemingly insoluble problem are really simple. We know that the sediments in the ocean once in solution are all ultimately derived from the igneous rocks; we also know, as the average result of numerous analyses of both igneous and detrital sedimentary rocks, that about 33 per cent. of the former pass into solution in the process of their conversion into the latter. Now of these substances going into solution, most are again precipitated from the ocean, or extracted by organic agency more or less rapidly, forming, in fact, the sub-oceanic sediments. The salts of sodium alone are neither precipitated nor extracted.[1] This is owing to their very high solubility. The recognition of this fact gives us a clue to many great generalizations in the history of solvent denudation. Among other facts, it serves to tell us what total amount of original igneous rock must have been broken up by denudation during the course of geological time.[2] This knowledge is obtained directly from the chemical analyses showing the average loss of sodium from igneous rocks when

[1] Joly, *Trans. Royal Dublin Society*, vol. vii., ser. ii., 1899, p. 23, *et seq.*

[2] Joly, *ibid.*, p. 46.

becoming sedimentary rocks, and the chemical analyses which give us the quantity of sodium in the sea.

Thus we find that the average igneous rock contains 3·39 per cent. of soda and the average sedimentary rock 1·30 per cent.[1] The loss has, therefore, been

$$3{\cdot}39 - 1{\cdot}30 \times \frac{67}{100} = 2{\cdot}52 \text{ per cent.}$$

in the process of conversion. Now the sodium in the ocean when converted into the equivalent oxide amounts to 21×10^{15} tonnes. The mass of the parent igneous rock is therefore :—

$$\frac{21 \times 10^{15}}{2{\cdot}52} \times 100 = 83{\cdot}3 \times 10^{16} \text{ tonnes.}$$

About 33 per cent. of this parent igneous rock passes into solution; and this amount would represent all the substances now in the sediments and still in solution in the ocean, but for the fact that some part of what has been precipitated in the ocean in past times is no longer to be reckoned among the deep-sea deposits, having been, in fact, elevated into dry land. Such are the limestones and some other deposits which, comparatively, are insignificant in amount. Van Hise[2] estimates the bulk of the calcareous rocks as amounting to about 5 per cent. of the total sedimentaries on the land. Now, as the insoluble detrital sediments are 67 per cent. of the total parent igneous rocks, there exist

$$83{\cdot}3 \times 10^{16} \times \frac{67}{100} = 55{\cdot}8 \times 10^{16}$$

tonnes of such detrital rocks, by far the larger part of

[1] Clarke, *The Data of Geochemistry*.

[2] Van Hise, *Treatise on Metamorphism*, p. 940. Somewhat higher estimates have been made by others.

which is on the land. According to Van Hise's estimate, then, the mass of detrital rocks plus the mass of calcareous rocks is obtained from the equation :—

$$M - \frac{M \times 5}{100} = 55{\cdot}8 \times 10^{16} \text{ tonnes};$$

from which $M = 58{\cdot}7 \times 10^{16}$ tonnes, and 5 per cent. of this is the amount of the calcareous rocks—that is, $2{\cdot}9 \times 10^{16}$ tonnes.

Hence, from 33 per cent. of the mass of the parent igneous rock—i.e., from $27{\cdot}5 \times 10^{16}$ tonnes—we must take $2{\cdot}9 \times 10^{16}$ tonnes, in order to allow for the limestones now elevated on the dry land. This leaves $24{\cdot}6 \times 10^{16}$ tonnes as the mass of the sediments on the floor of the ocean plus the mass of dissolved elements in the ocean. Taking the mass of the ocean as $1{\cdot}45 \times 10^{18}$ tonnes, and the dissolved substances therein as 3·5 per cent., the deduction for matter in solution is $5{\cdot}1 \times 10^{16}$ tonnes, leaving finally about $19{\cdot}5 \times 10^{16}$ tonnes. This is the mass of sediment precipitated on the ocean floor. The area of the ocean is estimated at 367×10^{12} square metres. If the mean specific gravity of such deposits as we are dealing with be 2·5, their bulk amounts to 8×10^{16} cubic metres nearly, and the average depth of precipitated sediments over the ocean floor would be about 218 metres, assuming that the rate of deposition had everywhere been alike throughout geological time. The latter supposition is, of course, imaginary.

It is probably safe to assume that the mean radioactivity of this vast quantity of sediment is no greater than that of Globigerina Oozes. Accepting

6×10^{-12} grams of radium per gram as the radioactivity, we arrive at the result that there are $1{\cdot}17 \times 10^{6}$ tonnes of radium, or well over one million tons, spread over the ocean floor. This is the record of the accumulated work of denudation throughout geological time: the quantity in the ocean waters is, as we have seen, relatively small. The result not only serves to bring before us a perspective view of the great phenomena with which we are dealing, but supports the finding on the sedimentary rocks, which appears to show that there is an appreciable difference in the radioactivity of the sedimentary and igneous surface-materials.

The detrital rocks can, as already observed, owe only a small part of their radium to the waters in which they have been deposited. The finer sediments of this class owe more to this source, probably, than those coarser ones which are practically composed entirely of discrete particles of older rocks. It is, of course, impossible to draw a hard and fast line between this group of rocks and those derived mainly by precipitation of dissolved substances. The following belong mainly to the detrital group:—

Welsh Slate (4), ...	5·8
Devonian Slate, Valencia, Co. Kerry, . .	3·6
Carbonaceous Shale, Burnlip Bore, Coatbridge, Fifeshire. Depth, 1173 ft., ...	8·7
Carbonaceous Shale, Burnlip Bore, Coatbridge, Fifeshire. Depth, 1168 ft., ...	6·7
Carbonaceous Sandstone, Burnlip Bore, Coatbridge, Fifeshire. Depth, 1095 ft., . .	2·9
Keuper Sandstone, ...	2·0

Argillaceous Shale, Balfour Bore, Fifeshire. Depth, 2124 ft., ... 3·6
Argillaceous Shale, Balfour Bore, Fifeshire. Depth, 2568 ft., ... 3·5
Argillaceous Shale, Balfour Bore, Fifeshire. Depth, 3420 ft., ... 3·7
Arkose Sandstone, Balfour Bore, Fifeshire. Depth, 3804 ft., ... 4·1
Carbonaceous Shale, Balfour Bore, Fifeshire. Depth, 4404 ft., ... 3·6
Quartz Grit, Balfour Bore, Fifeshire. Depth, 4530 ft., ... 5·1
Sandy Mud from bed of R. Amazon, Pará, . 3·2

The mean of these results is 4·4.

Some of these, as will be seen, come from considerable depths. The shales contain plant remains, and are probably of fresh-water origin.

The following belong to the second group of sedimentary rocks, and must have derived their radium mainly or entirely from the ocean:—

Limestone, Carboniferous, Burnlip Bore. Depth, 686 ft., 3·0
,, ,, Balfour Bore. ,, 2442 ft., 10·9
,, Ordovician, Trenton Valley, New York, . . 3·2
,, Carboniferous, Armagh, ... 3·4
,, Lias, Vaihingen, Wurtemberg, . . . 3·0
,, Solenhofen, Bavaria, ... 2·2
Dolomite, Zechstein, Osterhagen, Harz, ... 3·2
Chalk, Marsupites, ... 4·2
Coral Rock (recent), Funafuti Bore. Depth, 736–748 ft., 1·5
,, ,, ,, ,, 815–822 ft., 2·5
,, ,, ,, ,, 844–866 ft., 1·5
,, ,, ,, 874–936 ft., 1·3
Gypsum, Keuper, Westphalia, ... 6·9
Green Sandstone, Werl, Westphalia, ... 4·9

The mean of these results is 3·7.

Additional to the 27 rocks of the above tables may be quoted experiments on 19 rocks belonging to the sedimentary class and of Mesozoic age, from the St. Gothard tunnel. They are considerably altered by metamorphic action. They are cited more particularly further on. Again, 17 rocks of the Simplon tunnel which have been examined for radium are of the sedimentary class, and of Trias-Jura age. The means in each case are:—

Mesozoic Sediments, St. Gothard (19), . .	4·2
Trias-Jura Sediments, Simplon (17), . .	6·9

Thus a total of 63 rocks of sedimentary origin have been examined, giving the general mean of 4·7; or, if the Simplon results be excluded for reasons already given, the mean is 4·3.

Even when exceptionally high results are excluded, the average radioactivity of the igneous rocks, according to the present experiments, must be taken as approximating to 5 and of sedimentary rocks as near 4. For purposes of calculation further on, it has been thought well to take even lower values, partly because the general mean of all known results would fall below the above numbers, partly because we have to extend our estimates to materials which we cannot with certainty affirm to be represented on the surface.

It may be well to emphasize the fact that in selecting rocks for examination only unweathered specimens should be taken. Between the weathered rock and the final sandy residue containing but little radium it is only a matter of degree. Thus the sand of the Arabian Desert afforded but 0·4 as the radium content and in experiments made by other observers such substances as clays and soils invariably show the effects of weathering in removing the uranium and radium. Where our object is to find a value for the radium-content of the average crust-materials of the earth, it is evident that washed-out superficial residues, although in themselves of interest, should not be included.

Perhaps the most important problem of the future will be to find if any indication of diminishing radioactivity, as we go down in the crust, is to be detected. So far nothing definite on this point can be inferred. In parts the Deccan basalt shows a low radium-content; but the average is not lower than that of other rocks of the kind. The central St. Gothard rocks, as will later be seen, show the lowest radioactivities met with. There may be significance in this. But the whole question is at present hampered with the difficulty that we know but little as to the origin of the igneous rocks. We do not know the depths from which they have been brought; and it may be that their appearance at the surface is connected with a radioactivity greater than what generally prevails in the regions, beneath or within the crust, from which they originate.

The widespread occurrence of radium in the rocks affords an explanation of emanation not only in spring-waters, but in air drawn from soils and in the air of caves. Extending this conclusion, we see in the existence of gaseous substances among the series of radioactive elements a natural explanation of the radioactivity of the atmosphere, and, in part, of its state of ionization. The radioactive gaseous elements being in every known case substances of rapidly changing character, we recognize here the influence of a continual diffusion and convection of gases from the soils and rocks in making good the transformation of atmospheric emanation into solid forms. The experiments of Elster and Geitel[1] and others have shown that these solids may be collected by exposing negatively electrified wires to the air. Such electrified wires become, in fact, coated by active deposits which may be removed by solvents or by friction, as in the case of the emanations of thorium and radium. (See Chapter X.)

Again, there appears to be little doubt that the occurrence of helium in springs, in natural gases, and in the atmosphere (in which according to Ramsay it exists to the extent of four parts in a million) must be ascribed to the general diffusion of unstable elements in the surface-crust of the earth. For there is no longer any question of the identity of helium with the alpha ray. This subject is dealt with in another connexion (Chapter XI). Here it is sufficient to call attention to the confirmatory bearing of these facts upon the results of the investigations recorded in the foregoing pages.

The presence of thorium as a generally diffused rock-constituent has yet to receive the attention it deserves. Its emanation has been detected under various conditions both in the atmosphere and in underground waters;

[1] Elster and Geitel, *Phys. Zeit.*, ii., p. 590, and iii., p. 76, 1901.

and the experiments of Strutt[1] would appear to show that this element and its radioactive descendants are not without influence in supplying helium to the rocks.

It is possible that the aggregate radiothermal energy from this source may be not inconsiderable compared with that derived from the uranium family of elements.

A remarkable phenomenon by which the presence of radium may be detected by visual observation occurs in certain rocks. In brown mica (most generally biotite), in cordierite, and in some other minerals, small, circular, coloured spots, known as pleochroic haloes, have long been recognized by petrologists. These coloured areas are invariably found associated with a minute, centrally placed crystal, either of zircon, or more rarely of apatite.

The properties of the haloes are remarkable. In the case of biotite the medium in which the halo occurs is itself strongly pleochroic. A section of biotite taken transverse to the cleavage, when viewed by transmitted polarized light, darkens and brightens as the section is rotated over the polarized beam; the absorption being greatest when the plane of vibration of the ray is in the plane of cleavage of the mineral. A section which is taken parallel to the cleavage and viewed by transmitted polarized light shows only a feeble change of colour when turned round respecting the ray. The remarkably distinct cleavage properties of mica enable these sections to be readily distinguished under the

[1] Strutt, *Proc. R. S.*, vol. lxxv., pp. 88, 312; also vol. lxxx., p. 56.

microscope; the former showing fine longitudinal striations, the latter none.

A halo occurring in this medium shows all the above pleochroic properties intensified. Viewed in a section across the cleavage of the mica the halo may become quite opaque when the polarized ray vibrates in the plane of cleavage, and, again, may practically disappear when the vibrations are at right angles to this position; the central zircon being then most plainly defined and showing its characteristic strongly refracting properties. If the halo is observed on a mica section taken parallel to the cleavage, there is a comparatively slight change in the absorption when the ray or section is rotated.

It is evident that these haloes are really spherical in shape. For no matter in what direction the section is made, they still preserve a circular form, only departing from this if the central particle is elongated; in this case the halo, following the extension of the particle, becomes ellipsoidal: but when the zircon is a minute speck of roughly cubical form, the halo is found to be a circle of remarkable truth in every plane of section.

The mode of formation of these haloes has long been an enigma, and many investigations have been made in the hope of throwing light on their origin. It has been found that when the rock is highly heated the haloes disappear. Chiefly owing to this last observation, they have been ascribed by some petrologists to organic colouring matter.

A very little consideration should suffice to dispose of the view that they can be due to any diffused or aggregated colouring substance. For diffusion or aggregation could not possibly proceed at an equal rate across and along the cleavage of a medium possessing the structure of mica. In the plane of cleavage this mineral is a highly pervious substance, as is evinced by the intercalations of minerals, such as hematite, which have been introduced in solution, and also by the facility with which alterations take place in the direction of cleavage. The direction transverse to the cleavage is, on the other hand, one of remarkable impermeability, as every observation testifies. It is safe to affirm that no process of diffusion, whether gaseous or liquid, nor yet any process of aggregation, could take place in such a medium without being influenced by the difference of physical properties which the wonderful cleavage gives rise to.

Now, not only do we find that the haloes extend their boundaries uninfluenced by the crystalline cleavage, but cases are not uncommonly met with where the halo extends from one flake of mica to another quite uninfluenced by any mutual orientation of the flakes. This happens when the originating zircon is situated near the boundary of a mica crystal, and a second crystal of mica is sufficiently near to be affected. The photograph (Plate III.) shows such a case. The magnification is 300 diameters, and the light is polarized. The lower dark part of the halo is in a crystal of biotite so turned to the polarized rays that the absorption is nearly complete. The upper part of the halo, being in a differently oriented section of mica, shows a less degree of absorption. The central zircon is clearly

PLATE III.

Photograph of a pleochroic halo in a rock-section; magnified 300 diameters; light polarized. The halo-sphere extends across flakes of biotite mutually inclined, as is shown by the unequal absorption of the light.

seen. The rock here is a granite—a rock in which haloes are common.

In the case of the substance cordierite or iolite we are dealing with a medium of very different crystallographic properties from those of mica. There is not the remarkable cleavage, nor is the pleochroism conspicuous in thin sections. The substance possesses, however, pleochroic properties which in a small hand specimen are easily seen without the use of a polarizer; the crystal most generally appearing blue in one direction and straw-yellow in a perpendicular direction.

Haloes found in cordierite are not so conspicuous as in biotite. They are visible on certain sections of the mineral as pale-yellow stains surrounding the zircon enclosure. Rotated over the polarized rays, these stains brighten and darken.

There would certainly be less difficulty in ascribing the colouration in this case to the diffusion of a pigment. The colouration disappears when the rock is heated, as in the case of biotite.

That these haloes are caused by the radioactive properties of the centrally placed mineral particle, admits of little doubt.[1] The evidence may be summarized as follows:—(*a*) They seem to be invariably associated with the presence of a central substance of radioactive character. Some petrologists affirm that they occur with zircon only. But both zircon and apatite, according to the observations of Strutt,[2] are minerals of high radioactivity, containing, in every case examined, unusually large amounts of radium. Allanite is said to form,

[1] Joly, *Phil. Mag.*, March, 1907, p. 381. O. Mügge, *Centralbl. für Min.*, July, 1907.
[2] Strutt, *Proc. R.S.*, vol. lxxviii., p. 150.

sometimes, the central mineral: the author finds that this mineral is, in some cases at any rate, strongly radioactive.[1] (*b*) Theories of a pigment, etc., diffusing from the central particles are not tenable, as explained above. (*c*) There is exact correspondence between the radius of the halo and the distance to which the alpha radiation of radium would penetrate effectively, i.e., before its velocity is so reduced that its ionizing properties disappear. The halo appears never to exceed appreciably a certain maximum radial dimension of 0·05 millimetre. The average of a number of measurements was 0·04 millimetre. Smaller haloes are due generally to the section under observation being taken off the centre of the sphere, and, possibly, are occasionally due to a feeble source of radioactivity, so that the visible effect is confined to a region wherein the radiations are sufficiently dense. Rutherford has shown that thirteen layers of aluminium, each 0·00031 centimetre thick, reduce the velocity of the most energetic alpha radiations of radium to a degree rendering them incapable of effecting photographic actions or of exciting phosphorescence. The aggregate thickness of aluminium is here 0·04 millimetre, and the density of biotite (2·7 to 3·1) differing but little from that of aluminium (2·7), we see that the range of chemical activity of an alpha ray is in close correspondence with the radius of the halo. Such a coincidence should really be conclusive, for on any hypothesis of diffusion, etc., there need be no radial limits that we can define. (*d*) Colouration of like tint can be produced in cordierite[2] and biotite[3] by exposing

[1] Joly, *loc. cit.* [2] O. Mügge, *Centralbl. für Min.*, July, 1907.
[3] Joly, *Nature*, Oct. 10, 1907 p. 589.

these substances to the rays from specks of radium. The areas so coloured are pleochroic. In the case of mica the pleochroism is feeble, for the colouration thus artificially produced is viewed transverse to the cleavage surface, which in any case is a direction of feeble pleochroism.

It is true that the evidence from the experimental development of colouration is weakened by the probability that the change is here to be ascribed to the action of the more penetrating rays. Thus we may be dealing with the β and γ radiations which are known to colour many media. It is possible that the deep brown of certain micas much spotted by haloes may be due to these penetrating radiations. The coincidence of the radius with the effective range of the alpha ray, on the other hand, renders it practically certain that this particular radiation is responsible for the halo. It is probable that no true artificially formed halo has yet been produced. It may be added that Zirkel[1] records the occurrence of a grey marginal layer around the zircons of the Rhenish basalt. This layer is 0·05 millimetre thick, and is doubtless of similar origin to the halo. The examination of this layer for helium would be of interest.

What the halo is we do not know. Helium must be stored therein, but in very minute quantities. The presence of pleochroic haloes in rocks affords a visible proof of the sporadic and unequal distribution of radium to which we have already called attention.

[1] Zirkel, "Über Urausscheidungen in Rheinischen Basalten," p. 52.

CHAPTER IV.

UNDERGROUND TEMPERATURE AND RADIOACTIVITY.

UNSATISFACTORY STATE OF THE SUBJECT OF UNDERGROUND TEMPERATURE—ESTIMATES OF THE AVERAGE GRADIENT—IMPORTANCE OF THE GRADIENT—SIGNIFICANCE AND USE OF THE CONDUCTIVITY—THE CONDUCTIVITIES OF ROCKS—VARIATIONS OF CONDUCTIVITY WITH TEMPERATURE AND WITH CHANGE OF STATE—APPLICATION TO CALCULATE THE HEAT ESCAPING FROM THE EARTH—APPLICATIONS TO BORINGS AND TUNNELS—OUR KNOWLEDGE INSUFFICIENT—THE BALFOUR BORE—THE SIMPLON TUNNEL: TEMPERATURES ENCOUNTERED—EXPLANATIONS INADEQUATE—COMPARISON WITH TEMPERATURES IN THE CENTRAL ST. GOTHARD—CONDUCTIVITY OF THE SIMPLON ROCKS—THE RADIOACTIVITY OF THE SIMPLON ROCKS—DOWNWARD EXTENSION OF THE ROCKS REQUIRED TO ACCOUNT FOR THE ESCAPING HEAT—THE GEOLOGICAL STRUCTURE OF THE ST. GOTHARD—VARIATIONS OF GRADIENTS IN THE ST. GOTHARD—THE DISTRIBUTION OF RADIUM SUGGESTS AN EXPLANATION—THE GRADIENTS AND THE RADIOACTIVITY COMPARED.

THE subject of underground temperature is one of difficulty and obscurity. Much observational matter has been collected which can only be partially sifted from erroneous and misleading results, the pursuit of the subject being often left to those who are quite unqualified for the task. Observations are seldom made for their own sake, but under conditions which render them of secondary moment; whereas the subject is of such capital importance that investigations specially directed to the end in view would richly repay even large expense.

As the subject stands, there is grave doubt as to the reliability and significance of many of the published results, and much diversity of opinion exists as to what observations are specially deserving of weight. One fact appears clear, that, whether owing to local differences of conductivity or other causes, there is considerable variation in the recorded gradients. In mountain regions the circulation of water under gravity plays a considerable part, as the highly scientific observations made in the St. Gothard and Simplon tunnels have definitely proved. Sollas has suggested that in some cases the abnormal gradients may be associated with the distribution of molten matter beneath the crust.[1]

The following are some of the values, accepted by authorities on the subject, as fair averages of the results on the rise of temperature downwards:—

Prestwich, ...	1° C. in	24·3 metres
Lord Kelvin, ...	,,	27·5 ,,
Schardt,[2]	,,	32·0 ,,
Geikie,[3]	,,	27·0 to 32·4 metres
British Assoc. Committee, .	,,	32·4 metres
Clarence King, ...	,,	38·9 ,,

The mean of the values here collected is only a little above that adopted by the British Association Committee. Accordingly, in the applications which come later on, we take the average gradient as 32 metres for 1° C.

[1] Sollas, *The Age of the Earth*, p. 18.

[2] Schardt, *Verhandl. Schweizerischen Naturf. Gesellsch.*, 1904, 87, p. 204.

[3] Geikie, *Text-book of Geology*, vol. i., p. 64.

The importance of the gradient from the geological point of view is the information which, taken along with the conductivity, it affords as to the flow of heat to the earth's surface.

The reader who is not versed in the subject should endeavour to be quite clear on the connexion between the gradient and the heat-flow. The consideration of an analogous case may help. The flow of water through a uniform, narrow tube—that is, the number of cubic centimetres passing any cross-section per second—may be estimated if we know the difference of pressure at two points, say, one metre asunder; and also the effect of unit-pressure in driving water through unit-length of the tube. Let the pressures be measured by the equivalent head of water. Then if the difference of hydrostatic pressure at points one metre apart is, say, ten centimetres, and if we know by experiment that a difference of hydrostatic pressure of one centimetre at points one centimetre asunder, produces a flow of, say, five cubic centimetres per second, then we conclude that as the pressure gradient of ten centimetres per metre is one-tenth that which produces five cubic centimetres per second, the flow will only be the one-tenth part, or one-half a cubic centimetre per second.

Now the gradient of pressure in the case of the tube corresponds to the gradients of temperature we have already discussed as prevailing in the rocks; and the experimentally determined flow under the condition of unit-pressure gradient corresponds to the conductivity of the particular rock. In the latter case that which flows is heat, measured in so many calories or therms per second. When we know how many calories flow through a cross-section of one square centimetre for unit gradient—that is, a difference of temperature of 1° C. per centimetre—and also know the gradient prevailing in the rock, we have only to express the latter as such a fraction of a degree per centimetre, and then multiply by the conductivity, and we have the flow in calories per square centimetre. The calorie here is the quantity of heat required to raise one gram of cold water through one degree centigrade.

We see, then, that the indications of the gradient are dependent upon the conductivity of the particular class of rock in which the observations are made. Thus there might be cases in which the gradient was very different, but the flow of heat might be the same. Again, the gradients might be alike, but the flow of heat different. These principles will be required in connexion with various matters later on.

The thermal conductivity is again a very variable property; and we can only make general use of it by seeking some mean value. Rocks are what would be described as bad conductors. Under the conditions of gradient mentioned—i.e., 1° C. per centimetre—the flow in calories per square centimetre per second is in nearly every case some number in the third place of decimals.

The following are the most important determinations selected from the results published by the British Association Committee,[1] and in the Landholt-Börnstein

[1] Everett, *C.G.S. System of Units.*

Tabellen. The rocks are dry: moisture tends to increase the conductivity:—

Sandstone,	0·0055
Micaceous flagstone,	0·0053
Slate,	0·0048
Marbles, Limestone, and Dolomite,	0·0051
Caen stone,	0·0043
Chalk,	0·0025
Clay slate,	0·0027
Calcareous sandstone,	0·0021
Granite,	0·0053
Whinstone, Trap rock, and Mica-schist,	0·0038
Basalt,	0·0067
Syenite,	0·0044
Glass,	0·0017

The mean of these values is for the igneous materials 0·0042, and for the sedimentary 0·0041. For approximate calculations the average 0·004 sufficiently nearly represents the results on rocks generally.

A very important question is whether the conductivity increases or diminishes at high temperatures. There is insufficient information on the subject. R. Weber has found that gneiss, which has a conductivity of 0·00578 at 0°, falls in conductivity to 0·00416 at 100° C.;[1] and Lees has found, on the other hand, that the conductivity of glass increases by 0·0025 per cent. per degree between 35° and 60° C.[2] Metals appear to diminish in conductivity with rise of temperature; but alloys, on the other hand, improve in

[1] Landholt-Börnstein, *Tabellen*.

[2] Lees, *Phil. Trans.*, 1898, A, 399.

conductivity.[1] Such results as we possess on conductivity, traced through the solid and into the liquid state, would rather lead to the view that a fluid rock-magma must possess a lower conductivity still than the solid. Thus Lees found that ice falls in conductivity from the temperature of liquid air upwards, and shows a considerable diminution in the liquid state. Similarly in the case of aniline the conductivity solid is much greater than in the liquid state. In the case of glycerine the change is small, but still one of diminution.[2] Barus has obtained similar results in the case of thymol.[3]

Knowing the conductivity, it is easy to estimate the amount of heat which must annually escape at the earth's surface.

We have taken a mean gradient of 32 metres per degree, and a mean conductivity of 0·004 for the average rock. Expressing the gradient as the rise in temperature in a distance of one centimetre measured downwards, we have

$$\frac{1^\circ}{3200} \times 0{\cdot}004 = 1{\cdot}25 \times 10^{-6}$$

as the quantity of heat in calories which is conveyed to the surface per second per square centimetre of the earth's surface. If we multiply by the number of seconds in a year ($31{\cdot}5 \times 10^6$), and by the area of the earth in square centimetres (51×10^{17}), we finally arrive

[1] Lees, *Proc. R. S.* 80, A. 143. [2] *Ibid.*

[3] Barus, *Am. Journ. Sc.*, July, 1892.

at the total annual escape of heat from the earth. It comes out at $2 \cdot 0 \times 10^{20}$ calories.

Of course this must be regarded as only an approximation. The gradient beneath the oceans is unknown; we assume that it does not differ much from that which prevails (with variations) over the land surface. For the purpose of arriving at general views the result is probably close enough to the truth.

We have now to consider the principles which guide us in estimating the temperature effects produced by radium as a widely diffused rock-constituent.

We have already seen that the heat evolved by one gram of radium and its associated radioactive elements is closely $5 \cdot 6 \times 10^{-2}$ calories per gram of radium per second; that is to say, one calorie every eighteen seconds. This is, indeed, a rapid rate of evolution of energy from so small a quantity of matter. When, however, we come to deal with billionths of a gram of radium per gram of rock, of course the absolute quantity of heat evolved in any small volume of the rock is minute. On the other hand, in the case of the geological problems, there is practically unlimited time for its accumulation.

The accumulation is favoured by the nature of the circumstances. We are not dealing, it will be noticed, with an isolated quantity of radium deposited at a particular depth from the surface, but with materials containing radium disseminated more or less uniformly throughout their bulk. Hence the rock—whether igneous or sedimentary—is continually producing heat throughout every cubic centimetre of its volume. Thus, at any level beneath the surface, the heat produced in a particular cubic centimetre of the rock can only escape in virtue of this little mass of rock being hotter than the cubic centimetre of material immediately above it. But in this also heat is being developed, thus checking the escape of that which is developed beneath; and this reasoning applies all the way from base to top of the entire layer of rock: so that if no heat is escaping below, the maximum temperature must be at the base, and this basal temperature gives place to a lesser temperature as we go upwards—at first changing slowly, and then rather more rapidly near the surface.

When the question is treated mathematically, the equilibrium temperature θ, at any depth x from the surface, is found by the equation[1]:

$$\theta = \frac{qhx}{K}\left(D - \frac{x}{2}\right) \qquad (1)$$

where D is the total depth of the radioactive layer supposed neither to gain nor lose heat at the base, and to retain a constant temperature at the immediate surface; K is the conductivity of the particular rock; h is the constant of heat-production of radium per second; and q the amount of radium contained in one cubic centimetre of the rock.

At the very base of the layer, x becomes equal to D, and the basal or maximum temperature is

$$\theta = \frac{qh}{2K} D^2. \qquad (2)$$

[1] See Appendix A.

The first equation tells us that in order to determine the temperature at any particular depth in the layer, we must know the value of D or the total depth of the radioactive layer. This might indeed be evident from what we have already said about the temperature at every level influencing the temperature beneath, and, of course, in sending heat to the surface, influencing also the temperature above it. The second equation tells us the very important fact that the maximum temperature attained increases with the square of the depth of the radioactive layer. Thus, if equal quantities of radium were contained in two different radioactive strata, that stratum which possessed it most diffusedly distributed—that is which had less per cubic centimetre—would have the higher basal temperature; for the depth of the stratum is necessarily greater than that in which the radium is more concentrated. This, if at first appearing contradictory, will be understood when it is remembered that the general effect of the more diffuse distribution is to remove the radioactive materials further from the surface, and so enable the heat to be the better retained.

Applying these facts to the problem of interpreting the radioactivity of the surface-materials in terms of the gradient of temperature downwards, or, in other words, of endeavouring to connect the radioactivity with the underground temperature, we see that our deepest borings and tunnels are still too high up in the crust to enable us to pronounce with any certainty on the effect of radium met with in the rocks. The local radioactivity in itself can produce little effect on the temperature of the rocks.

An example will make this clear. The Simplon tunnel runs for a considerable distance at a depth, say, of 2,000 metres from the surface. The radioactivity of the rocks would appear to be about 19×10^{-12} grams per cubic centimetre, according to the experiments. Now, if we suppose that this radioactivity extends no further down than the level of the tunnel, the radium in the whole 2,000 metres would produce a temperature due to radioactivity of only 5·3° C. in the tunnel, this being the basal temperature of the layer.

We see from this how unlikely it is that volcanic temperatures can be brought very near the surface by the radioactivity of the rocks in the upper crust; for this involves a temperature of some 1200° being so produced. To bring about such a temperature, as a basal temperature, at a depth of, say, 7 kilometres from the surface, there must be a radioactivity throughout the 7 kilometres of over 340×10^{-12} grams per cubic centimetre. This is quite in excess of what prevails in any known considerable rock-mass. And as lavas, so far as examined, show no very excessive radioactivity, there can be no support for this theory of volcanic origin.

The gradients met with in deep borings, as already observed, vary considerably. The variations are doubtless in most cases due to variations in the rock-conductivity. But in some very erratic cases this may not be so, and it may be found possible to establish some connexion with the radioactivity of the local rocks. The majority of researches in this direction must, however, probably prove unfruitful. Consider in this connexion some radium measurements made in the case of the Balfour Bore in Fifeshire, which is carried through Carboniferous shales and sandstones to a depth of 4534·5 feet (1383 metres). The boring passes through a succession of beds changing frequently in character; the most continuous rock being a coarse, diabase-like layer some 500 feet thick. The shales contain vegetable fossils, and are probably of fresh-water origin. Throughout these materials, wherever tested, save in one case, there is a notable sameness in radioactivity:—

—	Depth from surface: feet.	Gradient: metres per degree Centigrade.	Radio-activity.
Black argillaceous rock, .	2124	45·3	3·6
Grey limestone, . .	2442	42·7	10·9
Argillaceous, fossiliferous,	2568	42·7	3·5
Whinstone, . . .	3000	42·7	3·8
Argillaceous rock, . .	3420	53·5	3·7
Arkose sandstone, . .	3804	34·0	4·1
Black shale, . .	4404	43·7	3·6
Quartz grit, . .	4530	43·7	5·1

It is probable that the varying gradients which prevail at the levels from which the samples are taken are referable almost entirely to the differing conductivities of the beds: the radioactivity only affecting the temperatures in quite a secondary manner.

In the case of the two great tunnels, the Simplon and the St. Gothard, there is evidence of the intervention of radioactivity in affecting the temperatures encountered in making the tunnels. This evidence is, however, only to be obtained by assuming—and the assumption is doubtless quite legitimate—the downward continuation of the radioactivity observed at the levels of the tunnels. The temperatures, which are so essential to any scientific discussion of the subject, were read in the rocks when the headings were being made. It has been found that the means taken to keep down the temperature in the progress of the works soon result in a fall of the rock-temperatures, so that the data to be of any real value must be obtained before this artificial source of cooling becomes effective. Nothing can exceed the pains devoted in both cases to this difficult work. Its value is incalculable, not only from the scientific but from the practical point of view.

The high temperatures met with in the Simplon, as all know, nearly put an end to the project, although the actual difficulty was more that arising from the *quantity* of heat which entered in the torrents of hot water than from the temperature of the rocks. The facts in outline are these:—For a stretch of some seven or eight kilometres the mean depth of the tunnel below the surface of the mountain is about 1700 metres. At the north end of this stretch, the rocks attain the temperature of about 55°, and this at the south end sinks to 35°. The temperature 55·4° was the highest encountered. It was not at the deepest part of the tunnel, but at a point somewhat to the north of it.

How unexpected such temperatures were, may be

shown by the predictions of the best authorities when the tunnel was still unattempted. Stapff—distinguished at once as engineer and as geologist—had in the case of the St. Gothard made a scientific study of the distribution of rock-temperature, and would rightly be regarded as the most competent authority of the time. His prediction for the Simplon was a maximum temperature of 47° C. He was, however, alone in making so high a prediction. Stockalper, also of St. Gothard fame, foretold 37° at a depth of 2000 metres from the surface. Heim thought 38° to 39° the probable maximum.[1]

These predictions were, doubtless, influenced by the accepted views as to gradients in mountains. The gradient actually attained at the hottest part of the tunnel would not be excessive for a boring made in the plains. Thus, if we take it that 50° C. was reached at 1600 metres below the mountain surface at 0° C., the gradient is no more than the average value of 32 metres per degree. However, previous experience of gradients in steep mountains would lead to the belief that the gradients in such cases are influenced by the convexity and steepness of the ground. In the case of the St. Gothard, the gradient for the major part of the tunnel was 46·6 metres,[2] and in the case of the Mont Cenis tunnel it has been estimated at 42·7 metres.[3] In

[1] Schardt, *Verhandl. Schweizerischen Naturf. Gesellsch.* 1904. 87, *Jahresversammlung*, p. 204.

[2] Stapff, *Trans. North of England Mining and Mech. Engineers*, xxxiii., 19 *et seq*.

[3] Prestwich, *Proc. R. S.*, xli., p. 1 *et seq*.

the last case the neglect of observations during the progress of the headings deprives the estimate of much of its reliability. Osmond Fisher, indeed, contends with much force that the convexity of the ground cannot exercise a marked influence on the gradients.[1] However, the fact remains that the temperatures observed must be regarded as above those generally experienced in previous cases, and, therefore, as abnormal on the evidence of the, perhaps, small number of observations of reliable character.

In explanation of the observed facts, it was pointed out that the north end of the tunnel, where the temperature was high, is exceptionally dry. The conductivity of the rock is, therefore, less, and there is no circulating water to carry off the heat; while the opposite conditions prevail at the south end. This is, of course, a physical explanation of why the gradients at the south end are low, and gives no reason for the high temperatures at the north end. The circulation of water is, in fact, a condition tending to disguise the true flow of heat at the south end if the average conductivity of rock is assumed where in truth it does not apply. The question really turns on the conductivity of the rock where the temperatures were highest; we want to know whether these temperatures indicate an abnormal heat-flow or not. The next explanation is, therefore, more to the point: the direction of the schistosity is appealed to. It is well known that considerable differences exist in the conductivity of certain rocks according as the heat-flow

[1] *Physics of the Earth's Crust*, 2nd ed., p. 222.

takes place across the schistose cleavage or along the cleavage. The explanation would be a good one if the schistosity was actually horizontal. In point of fact, the sections show it as pointing at a high angle to the horizontal at stretches near the north end where the temperature already nearly reached its maximum.[1] It cannot be doubted that the whole explanation of the very remarkable difference between the gradients in the central reaches of the Simplon and the St. Gothard tunnels is not fully explained in the angle of schistosity of the rocks in the former, nor yet, probably, in the conductivities, as we shall presently see.

In the case of the St. Gothard, the temperatures observed were of the most interesting character. In the central parts of the tunnel the mean gradient comes out as 46·6 metres per degree. These rocks are gneisses and schists of varying character. At the north end, where the tunnel passes through granite—the same granite that enters into the massif of the Finsteraarhorn—the gradient steepens to 20·9 metres. This last gradient, exceptionally steep even for a level plain, will again claim our attention. It is desirable to first confine our considerations to the temperature conditions where they are apparently least affected by special circumstances in the two mountains. Failing actual comparative measurements of the conductivities, we must assume that in the central St. Gothard there is actually a lesser flow of heat than prevails in the Simplon massif. We assume that in those stretches of both tunnels where there is no circulation of water

[1] Schardt, *loc. cit.*, and Schmidt, *Eclogie Geol. Helvetiæ*, ix., No. 484.

in the rocks, and both rocks are of much the same character, the differing gradient must represent a real difference in the quantity of heat escaping to the surface. In support of this assumption, it is to be observed that there appears no good reason to explain a difference in gradients of 32 and 46·6 metres, or nearly 50 per cent., on any quality of the rocks. Both are rocks of schistose and gneissic character, highly compressed, and not cooled by circulating water. The schistosity is more vertical, apparently, in the St. Gothard than in the Simplon; but this cannot create so great a difference in conductivity. Thus Zollinger[1] gives the conductivities in kilogram-calories per metre per hour of the schists and gneiss of the Simplon tunnel, taken across and with the cleavage, as 2·28 and 3·02 respectively; in the cases of calcareous schists and limestone, as 2·10 and 3·26; and in the case of granite 1·94 across the stratification, and 2·49 in the line of stratification. These reduced to the units previously used—the centimetre, gram, and second—give:—

Gneiss and schists across cleavage, . .	0·0063
,, ,, along ,, . . .	0·0084
Calcareous schists	0·0058
Limestone, ...	0·0090
Granite across stratification, ...	0·0054
,, along ,, ...	0·0069

As schistosity is oriented in the tunnel, only a fraction of the difference of conductivity in the several cases

[1] Discussion on Mr. Fox's paper: *Minutes, Proceedings Institute Civil Engineers*, 168, p. 53.

can be supposed to affect the flow of heat. It is also evident that there is nothing in the absolute values of the conductivities to account for the high temperatures in the Simplon; the conductivities are rather above the average values.

It would appear, however, that in the radium-content of the Simplon, contrasted with that of the St. Gothard, a possible explanation is forthcoming. The following summary embodies the results of experiments on the radioactivities of the Simplon rocks. A detailed statement of the individual results on the 49 rocks investigated is not necessary, and would hardly be desirable in view of the fact that a revision of the averages arrived at may be necessary (see p. 44).

Jurassic and Triassic altered sediments, . .	6·4
Crystalline schists, partly Jurassic and Triassic, partly Archæan, ...	7·3
Monte Leone gneiss and primitive gneiss, . .	6·3
Schistose gneiss—a fold from beneath, . .	6·5
Antigorio gneiss, ...	6·8

This subdivision of the Simplon rocks is that of Professor Schardt.

The mean of the individual experiments is $7{\cdot}1 \times 10^{-12}$ grams per gram, and if some exceptionally high readings be substituted for low ones obtained in the same rock, the mean rises to $9{\cdot}7 \times 10^{-12}$.

Now it will be presently seen that the experiments on the St. Gothard rocks do not give such high values unless in the granite at the north end of the tunnel. The gradient, according to Stapff,[1] declines from the

[1] Stapff, *loc. cit.*

PLATE IV.

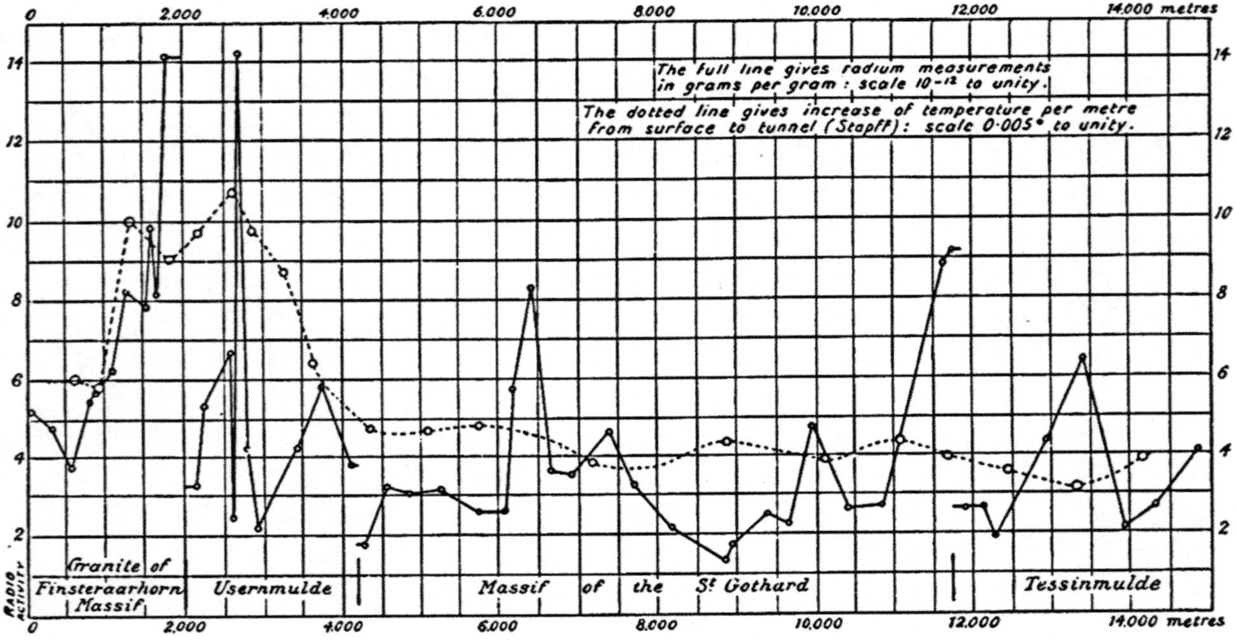

DISTRIBUTION OF RADIUM IN THE ROCKS OF THE ST. GOTHARD TUNNEL.

point 3550 metres measured from the north end to 4050 metres, after which it keeps a very low value to the south end of the tunnel. Water-circulation appears, however, to affect the gradients at the south extremity, giving rise to the lowest gradients anywhere observed in the tunnel. This applies probably to the last 3 kilometres. If we take the radium determinations from the beginning of the true St. Gothard massif to the distance 11,000 metres from the north end, we have the radium readings corresponding to the central low gradients. A glance at the accompanying chart, Plate IV., where the radium values are plotted as a full line, and the gradients as a dotted line, will more fully explain. The mean of these radium readings is 3·3. In this mean we have apparently a record of local radioactivity which we may compare, with security against causes of special disturbance, with the Simplon results. We find, in fact, the low radioactivity associated with the low gradients, and the high radioactivity with the high gradients.

We have already seen that without making assumptions as to the downward extension of the materials examined so near the surface, we cannot expect to connect our observations with the gradients. It is of interest to estimate what downward extension of the two rock-masses must be assumed to exist, always supposing them to maintain the observed mean radioactivities, in order to account for the temperature observations. Such an estimate is simple, if we suppose the conductivities alike throughout, and that therefore the differing gradients represent a different heat-flow to the surface. If now this difference of heat-flow is due to the different radioactivities of the rocks, then obviously there must be some distance to which if both masses extend the superior heat-flow from the Simplon becomes accounted for. This will be evident when it is remembered that each metre downward increases the superiority of the Simplon in respect of the total radium transmitting heat to the surface.

It has been explained that the product of the gradient and the conductivity is a quantity of heat flowing to the surface; accordingly, it follows that if R and r be the radium content per cubic centimetre of the Simplon and the St. Gothard rocks respectively, and G and g the gradients in each case, and K the conductivity, then when the rocks extend to a depth sufficient to produce the difference of heat-flow by the difference of total radium, we have $X(R-r)h = K(G-g)$, where X is the required depth, and h the heat-constant of radium. Inserting the values of the radioactivities per cubic centimetre (19×10^{-12}, and 8×10^{-12}), and taking K as 4×10^{-3}, and the gradients as $\frac{1}{3100}$ and $\frac{1}{4600}$ respectively, the value of X comes out as 7 kilometres. That is to say, a downward extension of 7 kilometres will establish the observed difference in gradient.

In the length of the St. Gothard tunnel a remarkable variation of gradient was found to exist between the northern sections and the central and southern parts. As already stated, the gradient near the north end is as high as 20·9 metres. In point of fact, a study of the rocks through which the tunnel is carried shows that the region in which this excessive gradient was found is geologically distinct from the St. Gothard massif. Entering at Goeschenen, the tunnel begins in the compact granite which is regarded as belonging to the Finsteraar massif. This granite, in parts gneissic, at a distance from the entrance of about 1800 metres, gives place to rocks of sedimentary origin, much altered, and of Triassic and Jurassic age. The sediments take the form of a downward fold or syncline, caught between the Finsteraar granite and the rocks of the St. Gothard massif proper. The downward extension of this synclinal fold cannot be great. It is petrographically continuous with the basin of the Ursern Thal, and is known as the Urserumulde. For two kilometres about, the tunnel traverses these altered sediments, which are of very varied character, before entering the highly metamorphosed rocks of the St. Gothard massif—rocks which appear to be really of unknown origin, but are generally regarded as igneous. Many varieties of schists and gneiss enter into this complex, which extends for about 7500 metres. The last section is run through the Tessinmulde for a distance of 3000 metres. These materials are again sedimentary in origin, and are mesozoic in age. (See Frontispiece.)

Stapff, in describing the variations in gradient in the northern, central, and southern parts of the tunnel, writes:—

"They [the isotherms] show irregularities on the south side, which clearly depend on cold springs; they bend down rapidly, and then run smoothly inclined beneath the water-filled section of the mountain.

Other local irregularities can be explained by the decomposition of the rock; but there is no obvious explanation of the rapid increase in the granite rocks at the northern end of the tunnel (2000 metres); and it is probably to be attributed to the influence of different thermal qualities of the rock on the coefficient of increase. For the rest these 2000 metres of granite belong to the massif of the Finsteraarhorn, and, geologically speaking, they do not share in the composition of the St. Gothard. Perhaps these two massifs belong to different geological periods (as supposed for geological reasons long ago). What wonder, then, if one of them be cooler than the other?"[1]

The suggestion here appears to be that some part of the original heat has been retained by the granite at the northern end of the tunnel. Prestwich modified this suggested explanation by referring the heat in the granite to the dynamical source arising from the movement of the Finsteraar granite into its existing position.[2]

The results of the examination of the rocks throughout the tunnel for radium seem to offer an explanation of less hypothetical character. It is only necessary to assume that the measurements of radium in the granite indicate the radioactive character of the rock in its downward extension beneath the in-folded Ursernmulde.

The chart showing the gradients and the measurements renders evident to the eye that at the north end

[1] Stapff, *loc. cit.*, p. 30. [2] Prestwich, *loc. cit.*, p. 44.

the radioactivity, which is not specially high at the entrance, increases rapidly after the first kilometre to 8 and ultimately to 14, just before the granite dips beneath the syncline. As already observed, the Ursern rocks can only go down a little way, and can therefore have but little influence on the supply of radioactive heat; and the gradients where the tunnel traverses these rocks must be referable to heat emanating from beneath them. Stapff shows in a detailed table of gradients beneath the plain of Andermatt that the gradients continue to rise to a point about 2500 metres from the north end, after which there is a rapid fall, so that when the St. Gothard rocks are met the gradient has fallen to nearly the low mean of the central section. Here the radioactivity of the rock changes to values about 3, rising, indeed, for short lengths, but again sinking, and ultimately reaching values only a little over 1. At the south end, where the syncline of the Tessin rocks meets the St. Gothard, a revival of radioactivity is indicated by the experiments; but in the Tessinmulde the radioactivities fall again. At this end of the tunnel the wetness of the rocks referred to by Stapff must remove the heat by convection, and no stress can be placed on the radioactivity.

Setting out the mean radioactivities of the different geological subdivisions of the tunnel, we have:—

Granite of Finsteraarhorn, ...	7·7
Sediments of Ursernmulde, ...	4·9
St. Gothard Massif, ...	3·9
Sediments of Tessinmulde, ...	3·4

The central section, however, if treated as a separate thermal province, would show a mean radioactivity of 3·3 only.

The measurements which yield these results were carried out after long experience on this work had been acquired. True it is that random specimens of the tunnel rocks would probably show here and there variations from the individual measurements; but unless a much larger number than fifty-one samples are dealt with, more representative information is improbable. A circumstance, quite accidental in character, which arose during the progress of the research is confirmatory of its representative character. A first set of observations was made on thirty-six rock-chips kindly given by the trustees of the British Museum. The chart so determined left many gaps, as well as readings which called for substantiation. Fifteen additional samples were obtained in order to complete, and, it was hoped, confirm, the chart. When the fresh samples were examined, it was found, in every case but one, that the results agreed with the previous neighbouring determinations, or, in other words, in one case only was any departure from the original chart necessitated. This is the reading 6·5 among the Tessin rocks. This seems evidence that accidental errors had not affected the experiments, and that the samples are representative in their indications.

It is unfortunate that we have no means of assuring ourselves as to the radioactive state of the granite beneath the Ursern rocks. But if in lieu of this knowledge we accept the last available reading in the granite as indicative of its radioactive state, and proceed as before to investigate the downward extension of rocks, having the radioactivities of 14 and 3·3, which will account for the differing heat-flows from the central and northern parts of the tunnel, we find that about 6 kilometres would suffice—the heat being conducted in an approximately vertical direction, and the conductivities of the two rock-masses being assumed alike. If some lesser radioactivity be assigned to the granite, so as to embrace a larger number of experiments, then the downward extension required, of course, becomes greater. It will be noticed that the downward extension arrived at for the central St. Gothard agrees with that arising from the comparison of the Simplon and St. Gothard rocks.

The foregoing pages give some idea of the difficulties and uncertainties attending this branch of research. There seems, however, to be a probability that a connexion between radioactivity and earth gradients has here been established. It is to be hoped that future work will enable us to pronounce upon this with certainty; for the matter is not only one of scientific interest, but of practical bearing on engineering problems of the nature and magnitude of that of the Simplon tunnel.

CHAPTER V.

THE INSTABILITY OF THE EARTH'S CRUST.

PHENOMENA OF MOUNTAIN-BUILDING—THEIR RHYTHMIC CHARACTER—PRIMARY CONDITION: THE COMPRESSIVE STRESS IN THE CRUST—THE BABBAGE-HERSCHEL EXPLANATION INADEQUATE—THE RADIO-THERMAL EFFECTS IN THE GREAT DEPOSITS—INCREASE OF THESE EFFECTS DUE TO THE RADIOACTIVITY OF THE UPPER CRUST—CALCULATIONS FOR SOME SPECIAL CASES—FURTHER RADIOACTIVE HEATING ARISING FROM LATERAL COMPRESSION—A RHYTHMIC CHARACTER NECESSARILY INVOLVED—THE GREAT GEOSYNCLINES TRACEABLE TO RADIOACTIVITY OF THE SEDIMENTS: SO ALSO THE ATTENDANT EARTHQUAKE AND VOLCANIC PHENOMENA—INFLUENCE ON GEOLOGICAL HISTORY.

IN the deposition of sediments—their uplifting into mountain chains—their subsequent removal and their re-deposition elsewhere—there is a continual convective movement of radioactive materials on the surface of the earth. Around the margins of all the great continents, where the sedimentation necessarily takes place to the greatest extent, the result of this convection is the accumulation of vast amounts of radioactive materials; and we have now to ask if this phenomenon may not entail consequences which can only be ascribed to the presence of the attendant radioactive energy.

The radioactivity of the secondary rocks is an assured fact arising out of the experiments undertaken to determine the amounts of radium in the surface materials of the earth. Whether ancient or modern, fresh or metamorphosed, the sediments in all cases exhibit radioactivity. Those that are detrital in origin owe their uranium directly to the parent igneous rocks; those which have been precipitated from the ocean take it from the waters. We are here in possession of a fact the consequences of which we are clearly entitled to pursue wherever the accumulation of these rocks finds place, and even if their aggregate growth amounts to depths measured in miles.

It has long been known to geologists that a mountain is not a mass of piled-up igneous rocks, but (save in the case of volcanic cones) is composed of a complex of sedimentary rocks ridged up, folded, and often horizontally over-thrust, by intense compressive forces. The great mountain ranges of the world, without exception, originated in the disturbance of sedimentary materials.

The mountain-forming movements have invariably taken place where the accumulation has been greatest. As we proceed away from the mountain we find the sediments rapidly thin out; and become relatively insignificant in depth where the continental surface remains undisturbed. This character appears, too, to have been persistent from the earliest times to the most recent, plainly revealing that in the fact of sedimentation the chain of events which results in the mountain is inaugurated.

The part played by igneous materials in these disturbances of the crust is only revealed by the discovering hand of denudation. When the foundations of the mountains are uncovered, we find that within, and like a supporting core, the underlying plastic rocks have also been upraised, following the upward movement of the sediments, and often injected into their folds, forced into cracks, and in many cases intensely metamorphosed by the heat, pressure, and shearing to which they were subjected.

Vast thicknesses of sediments are involved in these events. Indeed, whatever be the connexion between the deposition and upheaval, this element of great depth of accumulation seems a necessary condition, and must enter as a factor in the physical process involved. The mountain can only arise where the geosyncline is deeply filled by long ages of sedimentation.

The evidence gathered by geologists all over the world respecting the magnitude of these mountain-building sediments is uniformly impressive. There is, doubtless, the possibility of exaggerated measurements in some cases, but with every allowance the thicknesses are measured in miles. We might give up many pages to transcripts of such records; but a few facts must suffice to convey an idea of what is repeated, with little more than variations of locality, all over the earth. In the New World the broad structural features are, perhaps, more simple than in the Old. In the Proterozoic era of the North American continent the sediments were probably collected at a period when the crust of the earth was more readily disturbed than now. Three great sedimentary upheavals occurred: Huronian, Animikean, and Keweenawan; each followed by vast intervals of denudation, in which the upheaved crust was again worn down. To these intervals a duration as great as that of deposition has been ascribed. The removal of the sediments during these eras of denudation may perhaps be set off against errors affecting the estimates of thickness. Figures up to 18,000 feet have been cited for the depth of sediments originally taking part in the first of these systems, 14,000 feet for the second, and up to 50,000 feet for the third. Considerable amounts of igneous materials arising from beneath are associated with these accumulations—a feature which is, apparently, hardly so conspicuous in the sedimentary accumulations of more recent times. In later ages North American geological history moved on a still vaster scale. Dana,[1] in describing the fundamental characters of mountain-building phenomena, brings before us some of these great geological events in graphic language:—

"A mountain range of the common type, like that to which the Appalachians belong, is made out of the sedimentary formations of a long preceding era; beds that were laid down conformably, and in succession until they had reached the needed thickness; beds spreading over a region tens of thousands of square miles in area. The region over which sedimentary formations were in progress in order to make, finally, the Appalachian range, reached from New York to Alabama, and had a breadth of 100 to 200 miles, and the pile of horizontal beds along the middle was 40,000 feet in depth. The pile for the Wahsatch Mountains was 60,000 feet thick, according to King. The beds for the Appalachians were not laid down in a deep

[1] Dana, *Manual of Geology*, 3rd ed., p. 794.

ocean, but in shallow waters, where a gradual subsidence was in progress; and they at last, when ready for the genesis, lay in a trough 40,000 feet deep, filling the trough to the brim. It thus appears that epochs of mountain-making have occurred only after long intervals of quiet in the history of a continent."

On the western side of North America the work of mountain-building was, indeed, on the grandest scale. For long ages and through a succession of geological epochs sedimentation had proceeded so that the accumulations of Palæozoic and Mesozoic times had collected in the geosyncline formed by their own ever increasing weight. The site of the future Laramide range was in late Cretaceous times occupied by some 50,000 feet of sedimentary deposits; but the limit had apparently been attained, and at this time the Laramide range, as well as its southerly continuation into the United States, the Rockies, had their beginning. Chamberlin and Salisbury[1] estimate that the height of the mountains, developed in the Laramide range at this time was 20,000 feet, and that, owing to the further elevation which has since taken place, from 32,000 to 35,000 feet would be their present height if erosion had not reduced them. Thus on either side of the great continent we have the same forces at work, throwing up mountain ridges where the sediments had formerly been shed into the ocean.

In the Himalayas an immense mass of sediments, laid down on the floor of the ancient Tethys, was at the

[1] Chamberlin and Salisbury, *Geology*, 1906, iii., 163.

end of the Eocene period first disturbed. Conditions of quiescent deposition since the Upper Carboniferous thus came to an end, and a succession of mountain-building movements was inaugurated in North India—the final movements extending into Pliocene times and the sedimentation continuing over those areas not yet uplifted above the sea-level. The masses of sediment involved in these prolonged eras of deposition and partial upheaval are now, in view of the almost incredible erosion which has since occurred, hardly determinable, but must be enormous. The Alps and the Pyrenees share in this history. Nor would the list of such crustal movements be complete save in the enumeration of every great range upon the earth.

These great events are of a rhythmic character; the crust, as it were, pulsating under the combined influences of sedimentation and denudation. The first involves downward movements under the gathering load, and ultimately a reversal of the movement to one of upheaval; the second factor, which throughout has been in operation as creator of the sediments, then intervenes as an assailant of the newly raised mountains, transporting their materials again to the ocean, when the rhythmic action is restored to its first phase, and the age-long sequence of events must begin all over again.

It has long been inferred that compressive stress in the crust must be a primary condition of these movements. The work required to effect the upheavals must be derived from some pre-existing source of energy. The phenomenon—intrinsically one of folding of the crust—suggests the adjustment of the earth-crust to a lessening radius; the continual loss of heat from the surface of the earth affording an explanation of this shrinkage, and the assumption of compressive stress in this folding envelope naturally arising as a harmonious part of the general theory. The compressive stresses must, however, be confined to the upper few miles of the crust, for, in fact, the downward increase of temperature and pressure must soon confer fluid properties on the medium, and slow tangential compression must result in hydrostatic pressure rather than directed stresses. Thus the folding visible in the mountain range, and the lateral compression arising therefrom, are effects confined to the upper parts of the crust.

The energy, then, which uplifts the mountain is a surviving part of the original gravitational potentional energy of the crust itself. It must be assumed that the crust in following downwards the shrinking subcrustal magma, as this lessens its volume in the process of congelation and crystallization, develops immense compressive stresses in its materials, vast geographical areas being involved. When folding at length takes place along the axis of the elongated syncline of deposition, the stresses find relief probably for some hundreds of miles, and the region of folding now becomes compressed in a transverse direction. As an illustration, the Laramide range, according to Dawson, represents the reduction of a surface-belt 50 miles wide to one of 25 miles. The marvellous translatory movements of crustal folds from south to north arising in the genesis of the Swiss Alps, which recent research has brought to light, is another example of these movements of relief, which continue to take place perhaps for many millions of years after they are initiated.

The result of this yielding of the crust is a buckling of the surface which on the whole is directed upwards; but depression also is an attendant, in many cases at least, on mountain upheaval. Thus we find that the ocean floor is depressed into a syncline along the entire western coast of South America; a trough always parallel to the ranges of the Andes. The downward deflection of the crust is of course an outcome of the same compressive stresses which elevate the mountain.

The fact that the yielding of the crust is always situated where the sediments have accumulated to the greatest depth, has led to attempts from time to time at establishing a physical connexion between the one and the other. The best-known of these theories is that of Babbage and Herschel. This seeks the connexion in the rise of the geotherms into the sinking mass of sediment and the consequent increase of temperature of the earth-crust beneath. It will be understood that as these isogeotherms lie at a uniform distance from the surface of the earth, unless where special variations of conductivity may disturb them, the introduction of material pressed downwards from above must result in these materials partaking of the temperature proper to the depth to which they are depressed. In other words, the geotherms rise into the sinking sediments, always, however, preserving their former average distance from the surface. The argument is that as this process undoubtedly involves the heating up of that portion of the crust which the sediments have displaced downwards, the result must be a local enfeeblement of the crust, and hence these areas become those of least resistance to the stresses in the crust.

When this theory is examined closely, we see that it only amounts to saying that the bedded rocks, which have taken the place of the igneous materials beneath, as a part of the rigid crust of the earth, must be less able to withstand compressive stress than the average crust. For there has been no absolute rise of the geotherms, the conductivities of both classes of materials differing but little. Sedimentary rock has merely taken the place of average crust-rock, and is subjected to the same average temperature and pressure prevailing in the surrounding crust. But are there any grounds for the assumption that the compressive resistance of a complex of sedimentary rocks is inferior to one of igneous materials? The metamorphosed siliceous sediments are among the strongest rocks known as regards resistance to compressive stress; and if limestones have indeed plastic qualities, it must be remembered that their average amount is only some 5 per cent. of the whole. Again, so far as rise of temperature in the upper crust may affect the question, a temperature which will soften an average igneous rock will not soften a sedimentary rock, for the reason that the effect of solvent denudation has been to remove those alkaline silicates which confer fusibility.

When, however, we take into account that in the radium content of the sediments there is a source of heat competent to bring the geotherms closer to the surface in the region of deposition than elsewhere in the surrounding crust, the difficulties disappear. We can calculate what these heating effects will amount to. Let us take first a simple case: a succession of beds such as Dana refers to, 40,000 feet in thickness—that is, somewhat over 12 kilometres. Looking at the results already recorded for sedimentary materials, we see that 4×10^{-12} grams per gram is not too large an estimate of the average radioactivity. This is 11×10^{-12} per cubic centimetre. Then we find for the basal temperature of the layer of sediment (see p. 77):—

$$\theta = \frac{11 \times 10^{-12} \times 5.6 \times 10^{-2}}{2 \times 4 \times 10^{-3}} \times 144 \times 10^{10} = 111^\circ \text{ C.}$$

That is, there will be an elevation of about 111° above that which similar materials under the same conditions, but without radioactive ingredients, would experience. Thus the final position of the isogeotherms is no longer alike in the sediments and elsewhere in the crust, as necessarily arises in the Babbage-Herschel theory, but there has been an upward movement due to radioactivity. If the gradients be supposed similar to what they now are—say, 30 metres to 1 degree—the upward shift has been 3330 metres at the base of the layer.

It would be quite comprehensible that even this amount of heating would appreciably weaken the rigid part of the crust over this area. For the heating effect will extend downwards, and the temperature will rise a corresponding amount in the depressed underlying parts. To make this more apparent, suppose there was as much as 40 kilometres of the crust rigid: then the geotherm—say of 800°—originally defining the downward limit of rigidity, moves up 3 kilometres, and the thickness of the rigid part falls to 37 kilometres, and there is, consequently, a reduction of 7½ per cent. in the effective resistance, supposing all other factors alike. When the compressive stress in the crust over this region of the earth, increasing with the lapse of time, attains to nearly the limiting resistance, the flexure would, almost certainly, begin here, and once begun, there would be not only mechanical reasons for its continuance, but accelerative radioactive effects would arise from a source to be referred to later. Thus the deposition of the sediments would, under the conditions of a shrinking world, involve their own upheaval.

The problem of the influence of radium in mountain elevation is, however, not so simple as given above: the actual effects produced are at once more striking and more complex. It will be shown in a later chapter that there must be over the earth's surface a specially rich radioactive layer, and that such radioactivity as exists beneath it must extend downwards in materials relatively poor in uranium. The depth of the uranium-rich upper part of the crust can be guessed at from the necessity of reconciling its properties with certain facts; and the result of our guess is that it must extend to a depth of between 10 and 15 kilometres. Along with the existence of this layer the important fact must be held in view that the basal temperature of any radioactive layer increases as the square of its thickness. Thus the effect of the sediment collecting upon the normal radioactive covering of the earth will be to produce a total thickness of radioactive materials — made up of the normal layer and of the sediments—greater than exists elsewhere, and the temperature at the base of the underlying normal layer will be so much greater than it was before, as the square of the combined thickness of layer and sediments is to the square of its thickness elsewhere. That is, if the thickness of the normal layer is, say, 12 kilometres, and its basal temperature so far as due to radium is t_1, and sediments are laid down upon it, making it, say, 22 kilometres, the basal temperature t_2 is obtained by the proportion $t_1 : t_2 :: 12^2 : 22^2$. We assume the conductivity and radioactivity of the sediments the same as prevails in the normal layer.

The last assumption is not completely justified on the experiments. The error of assuming the radioactivities to be alike will, however, be unimportant. In any case we will take a value for the common radioactivity considerably below the mean of all existing experiments, mindful of the fact that we really do not know how far down the surface radioactivity is carried at its full value. Judging from the indications of igneous rocks brought to the surface, probably from considerable depths, there is no reason to suppose a falling off. We accept, however, for our calculations the radioactivity $3{\cdot}0 \times 10^{-12}$ grams per gram, or $8{\cdot}4 \times 10^{-12}$ grams per cubic centimetre for the entire complex layer made up of sediments and the underlying radioactive materials. The precise figure we adopt in our calculations is of secondary importance, for the margin is wide within which we may suppose it to vary (according as future researches may decide), and the deductions now to be considered still retain their fundamental significance.

Let us consider the cases of sedimentary deposits amounting to 4, 6, 8, 10, 12, and 14 kilometres. To these thicknesses we add the inferred depth (12 kilometres) of the normal layer. Here again it must be observed that wide variations in our assumption will not seriously affect the result. The depth taken above for purposes of calculation has been arrived at on a basis distinct from the present investigation. It is certain that we may assume a lesser or a greater depth, varying it through a wide range, and still the teaching of the result will remain the same.

The total depths are, then, 16, 18, 20, 22, 24, 26 kilometres. We take the average conductivity as 4×10^{-3}. It will now conduce to clearness if we follow the calculation in the case of one of these assumed depths; recollecting that all present the same problem.

We easily find on the equation for θ (p. 77) that the temperature due to radioactivity at the base of the 22-kilometre layer (which we select) is 285°. Now what we want to find is the weakening of the crust arising from this radiothermal temperature: that is, how much weaker the crust is here than it is elsewhere. In other words, the first consideration is:—

what is the thermal state of the crust, assumed practically free from overlying sediment, elsewhere at this depth? The temperature generally prevailing at 22 kilometres involves, of course, the gradients downwards from the surface. Here we recall the mean gradient of 32 metres, assuming that the temperature conditions are like those of to-day. This is, however, the observed *surface* gradient, and we know that this surface gradient cannot apply to the deeper parts of the crust, owing to the disturbing effects of the radium in the normal layer. Below this layer the gradient must be less. We desire to distinguish between the heating effects of the radium near the surface and the heat due to conducted interior heat. The simplest way of investigating this—and it is accurate enough for our purpose—is to deduct the heat out-put of the normal layer per square centimetre of the earth's surface from the average heat-emission of the earth. A simple calculation shows that $0 \cdot 564 \times 10^{-6}$ must be deducted from the $1 \cdot 250 \times 10^{-6}$ calories previously arrived at (p. 75) as the out-flow per square centimetre per second. The balance of $0 \cdot 686 \times 10^{-6}$ gives us the interior heat-gradient (p. 73 *et seq.*). It comes out as 1 in 58·4 metres or, say, 58 metres.

We are now in a position to determine the temperature of the geotherm at places unaffected by sedimentation at the depth of 22 kilometres. The radium in the upper 12 kilometres creates a basal rise of nearly 85°, which must affect all points beneath this level. The out-flowing interior heat gives rise to 379° as calculated on the gradient of 58 metres;

and therefore the temperature at the depth of 22 kilometres must be 379° + 85°, that is 464°.

Returning now to the new conditions, when, of the 22 kilometres, 10 are composed of sediment and 12 of the normal radioactive layer, we find a radium temperature of 285°, and to this must be added the temperature due to the interior heat. At this level we have found this to be 379°. The total temperature is therefore 664°. We have thus found that the effect of the radioactive stratum of sediments placed on the normal radioactive layer has been to raise the temperature at the depth of 22 kilometres (where the rise will have reached the maximum) from 464° to 664°; that is, by 200°.

This, of course, is an upward displacement of the geotherms. By how much are they displaced? The slope or gradient of temperature is at points deep down 1 in 58 metres, and the temperature, at the depth of 22 kilometres, has been raised 200°. This increase would only be gained elsewhere by going down 11·6 kilometres: and hence this has been the elevation of the geotherms beneath the radium-rich materials.

This calculation is made for each of the sedimentary accumulations already selected. We can best compare and estimate the effects of the different depths of deposit by finding the upward shift of some one geotherm. We take that having the temperature of 800°, assuming that this approximately represents the softening temperature of the crust rocks. Possibly a owever temperature would be better chosen. Now, normally this geotherm lies at about 40 kilometres down, and we find, finally, the results:—

Thickness of Sedimentary Deposit.	Resulting Rise of Isogeotherms.	Weakening of Earth's Crust as defined by the Rise of the Geotherm at 40 kilometres.
Kilometres.	Kilometres.	Kilometres.
4	3·8	40 to 36·2
6	6·2	40 to 33·8
8	8·7	40 to 31·3
10	11·6	40 to 28·4
12	14·8	40 to 25·2
14	18·2	40 to 21·8

Looking at these figures we see that even the 4-kilometre deposit involves a reduction of the effective strength of the earth's crust by 10 per cent.

Such results are, of course, approximate only. More cannot be claimed for them than that they represent roughly the magnitude of the effects it is sought to investigate. They show that as the sediments grow in depth there is a rising of the geotherm of plasticity—whatever its true temperature may be—gradually reducing the thickness of that part of the upper crust which is bearing the simultaneously increasing compressive stresses. Below this geotherm long-continued stress resolves itself into hydrostatic pressure; above it (there is, of course, no sharp line of demarcation) the crust accumulates elastic energy. The final yielding and flexure occur when the resistant cross-section has sufficiently diminished. It is probable that there is also some outward hydrostatic thrust over the area of rising temperature, which assists in determining the upward throw of the folds.

When yielding has begun in any geosyncline, and the materials are faulted and overthrust, there results a considerably increased thickness. As an instance, consider the piling up of sediments over the existing materials of the Alps, which resulted from the compressive force acting from south to north in the progress of Alpine upheaval. Schmidt of Basel has estimated that from 15 to 20 kilometres of rock covered the materials of the Simplon as now exposed, at the time when the orogenic forces were actively at work folding and shearing the beds, and injecting into their folds the plastic gneisses coming from beneath.[1] The lateral compression of the area of deposition of the Laramide, already referred to, resulted in a great thickening of the deposits. Many other cases might be cited; the effect is always in some degree necessarily produced.

If time be given for the heat to accumulate in the lower depths of the crushed-up sediments, here is an additional source of increased temperature. The piled-up masses of the Simplon might, even on a low estimate of the radioactivity, have occasioned a rise due to radioactive heating of from 300° to 500°; and if this be added to the interior heat, a total of from 800° to 1100° might have prevailed in the rocks now exposed at the surface of the mountain. Even a lesser temperature, accompanied by the intense pressure conditions, might well occasion the appearances of

[1] Schmidt, *Ec. Geol. Helvetiæ*, vol. ix., No. 4, p. 590.

thermal metamorphism described by Weinschenk, and for which, otherwise, there is difficulty in accounting.[1]

This increase upon the primarily developed temperature conditions takes place concurrently with the progress of compression; and while it cannot be taken into account in estimating the conditions of initial yielding of the crust, it adds an element of instability, inasmuch as any progressive thickening by lateral compression results in an accelerated rise of the geotherms. It is possible that time sufficient for these effects to develop, if not to their final, yet to a considerable extent, was available. The figures in the case of the Simplon plainly show that rock-melting temperatures would be brought about at but a little distance below the present surface. The viscous movements of siliceous materials, and the out-pouring of igneous rocks which often attend mountain elevation, would find an explanation in such temperatures.

There is no more striking feature of the part here played by radioactivity than the fact that the rhythmic occurrence of depression and upheaval succeeding each other after great intervals of time, and often shifting their position but little from the first scene of sedimentation, becomes accounted for. The energy, as we have already remarked, is in fact transported with the sediments—the energy which determines the place of yielding and upheaval, and ordains that the mountain ranges shall stand around the continental borders. Sedimentation from this point of view is a convection of energy.

[1] Weinschenk, *Congrès Géol. Internat.*, 1900, i., p. 332.

When the consolidated sediments are by these and by succeeding movements forced upwards into mountains, they are exposed to denudative effects greatly exceeding what affects the plains. Witness the removal during late Tertiary times of the vast thickness of rock enveloping the Alps. Such great masses are hurried away by ice, rivers, and rain. The ocean receives them; and with infinite patience the world awaits the slow accumulation of the radioactive energy beginning afresh upon its work. The time for such events appears to us immense, for millions of years are required for the sediments to grow in thickness, and the geotherms to move upwards; but vast as it is, it is but a moment in the life of the parent substance, whose atoms, hardly diminished in numbers, pursue their changes while the mountains come and go, and the rudiments of life develop into its highest consummations.

The broad features of the denuded earth are grand in their simplicity. A geosyncline borders the Pacific, in past times the recipient of vast sedimentary accumulations. It follows the wall of mountains standing up on the western side of North and South America for a distance of some 8000 miles. It reappears at New Zealand, and passing north of Australia contains the festoons of islands which skirt Malaysia and eastern Asia. A branch diverging to the west lies along the equator, and bends northward, taking in the Himalayas, the Alps, and the Pyrenees. In short, the geosynclines lie where the oceans meet the continents, or did so in the past, and conserve the sediments near the land.

They embrace the greatest mountain ranges on the earth, and now hold the greater part of the material eroded from the land throughout the long ages of Mesozoic denudation.

It seems impossible to escape the conclusion that in this great development of ocean-encircling areas of deposition and crustal folding, the heat of radioactivity plays a part. The geosynclines were a necessary outcome of the permanency of the continents, and originated in the age-long accumulation of the denuded radioactive energy. We recognize, therefore, in the movements of the sediments not only an influence localizing and accelerating crustal movements, but one which, in subservience to the primal distribution of land and water, has determined some of the greatest geographical features of the globe.

It is no more than a step to show that bound up with the radioactive energy are most of the earthquake and volcanic phenomena of the earth. The association of earthquakes with the great geosynclines is well known. The work of De Montessus showed that over 94 per cent. of all recorded shocks lie in the geosynclinal belts. There can be no doubt that these manifestations of instability are the results of the local weakness and flexure which originated in the accumulation of energy denuded from the continents. Similarly we may view in volcanoes phenomena referable to the same fundamental cause. The volcano was, in fact, long regarded as more intimately connected with earthquakes than it, probably, actually is; the association being regarded in a causative light, whereas the connexion is more that of possessing a common origin. The girdle of volcanoes around the Pacific and the earthquake belt coincide. Again, the ancient and modern volcanoes and earthquakes of Europe are associated with the geosyncline of the greater Mediterranean, the Tethys of Mesozoic times. There is no difficulty in understanding in a general way the nature of the association. The earthquake is the manifestation of rupture and slip, and, as Suess has shown, the epicentres shift along that fault-line where the crust has yielded.[1] The volcano marks the spot where the zone of fusion is brought so high up in the fractured crust that the melted materials are poured out upon the surface.

In a recent work on the subject of earthquakes Professor Hobbs writes :—" One of the most interesting of the generalizations which De Montessus has reached as a result of his protracted studies, is that the earthquake districts on the land correspond almost exactly to those belts upon the Globe which were the almost continuous ocean basins of the long Secondary era of geological history. Within these belts the sedimentary formations of the crust were laid down in the greatest thickness, and the formations follow each other in relatively complete succession. For almost or quite the period of this long era it is therefore clear that the ocean covered these zones. About them the formations are found interrupted, and the lacunæ indicate that the sea invaded the area only to recede from it, and again at some later period to transgress upon it. For a long time, therefore, these earthquake belts were the sea

[1] Suess, *The Face of the Earth*, vol. ii., chap. ii.

basins—the geosynclines. They became later the rising mountains of the Tertiary period, and mountains they are to-day. The earthquake belts are hence those portions of the earth's crust which in recent times have suffered the greatest movements in a vertical direction —*they are the most mobile portions of the earth's crust.*"[1] The italics are Professor Hobbs's.

Whether the movements attending mountain elevation and denudation are a connected and integral part of those wide geographical changes which result in submergence and elevation of large continental areas, is an obscure and complex question. We seem, indeed, according to the views of some authorities, hardly in a position to affirm with certainty that such widespread movements of the land have actually occurred, and that the phenomena are not the outcome of fluctuations of oceanic level; that our observations go no further than the recognition of positive and negative movements of the strand.

However this may be, by far the greater part of denudation during geological time has been done on the mountain ranges. There is apparently an accelerative principle involved in this fact, tending to increase the magnitude of mountain ranges with the advance of geological time. It is, in short, indisputable that the orogenic movements which uplift the hills have been at the basis of geological history. To them the great accumulations of sediments which now form so large a part of continental land are mainly due. There can be no doubt of the fact that these move-

[1] Hobbs, *Earthquakes*, p. 58.

ments have swayed the entire history, both inorganic and organic, of the world in which we live.

To sum the contents of this chapter in the most general terms, we find that in the conception of denudation as producing the convection and accumulation of radio-thermal energy the surface features of the globe receive a new significance. The heat of the earth is not internal only, but rather a heat-source exists at the surface, which, as we shall see, cannot prevail to the same degree within; and when the conditions become favourable for the aggregation of the energy, the crust, heated both from beneath and from above, assumes properties more akin to those of its earlier stages of development, the secular heat-loss being restored in the radioactive supplies. These causes of local mobility have been in operation, shifting somewhat from place to place, and defined geographically by the continental masses undergoing denudation, since the earliest times.

CHAPTER VI.

THE INSTABILITY OF THE OCEAN FLOOR.

DARWIN'S THEORY OF CORAL-ISLAND FORMATION—EXTENT OF UNSTABLE AREAS—CONDITIONS OF DEEP-SEA DEPOSITION—MURRAY'S THEORY—FUNAFUTI—THE SEDIMENTS AROUND CORAL ISLANDS—THE LAWS OF SEDIMENTATION: THEIR GEOLOGICAL IMPORTANCE—THE RADIOACTIVE STATE OF THE SEDIMENTS—INCREASE OF RADIOTHERMAL EFFECT WITH DILUTION—TIME OF ACCUMULATION—PERMANENCY OF THE OCEAN BASINS—THE EXISTENCE OF SUBMERGED CONTINENTS INCONSISTENT WITH THE DATA OF SEDIMENTATION—RATE OF DEPOSITION OF DEEP-SEA SEDIMENTS—THE BARBADOS DEPOSITS—RATE OF COLLECTION OF THE CHALK—EVIDENCE OF VARIATION IN RATE OF ACCUMULATION—CONCLUSION AS TO POSSIBLE WEAKENING OF THE CRUST BENEATH OCEANIC DEPOSITS—EFFECTS OF THE INSTABILITY ON THE MEAN SEA-LEVEL AND THE TRANSGRESSION OF THE OCEAN UPON THE LAND.

IN the case of mountain elevation as discussed in Chapter V, the conditions necessary for the accumulation of radioactive heat undoubtedly exist, and temperatures adequate to affect the local stability of the crust must be attained. The whole theory is of the simplest kind. Given a stressed crust and a local source of heat applied above while the normal heat of the earth flows upward from beneath, and the area where these conditions exist must necessarily become the first place of yielding and flexure; as naturally as the rupture of the chain occurs at the weakest link. Under these conditions accordingly, we find associated with great accumulations of radioactive sediments bending and fracture of the crust, with the attendant phenomena of earthquake and volcanoes. The cause and the result are both before us.

There exist, however, on the surface of the earth areas where indeed the results of instability—restlessness and vulcanicity—appear evidently displayed, but the demonstration of the adequate cause is at present denied to us, and possibly may always be so. It is of interest to examine how far conjecture as to the cause in these cases may be guided by the knowledge at our disposal.

Darwin long ago urged that in the facts connected with the coral islands or atolls there is evidence of instability of the earth's crust. On the hypothesis of downward movement he proposed to reconcile the existence of these steeply ascending structures of coral with the limitation which confines the operations of the reef-building polyps to depths of some fifteen or twenty fathoms of water. The volcanic cone, then doubtless active and casting its ashes around, had first to rise above or near to the surface of the sea. Later this must sink, affording, however, a foundation suitable to the needs of the reef-builders. And the patient architects, ever keeping pace with the downward movement of the cone, had maintained at the sea's surface a visible monument of the volcano which once had been.

Admittedly, Darwin's theory involves movements of the ocean floor in many places and over areas of great extent. Nor are the movements downwards only, but also upwards, where we see the once submerged coral standing high above sea-level. We must conclude that waves of flexure or of buckling affected the earth's crust over these oceanic areas for long periods of time, but periods which probably should be referred to the comparatively recent past.

The areas involved are really enormous. The south-equatorial coral region extends nearly 7000 miles in latitude, and from 1000 to 1500 miles in longitude—an area about three times as great as that of Australia. The Maldive Archipelago of coral islands in the Indian Ocean is relatively small—about 23,000 square miles. These and other coral islands exhibit everywhere the same characteristics. They everywhere emerge not from the deepest ocean, but from plateaux seldom or never over 2000 fathoms beneath the surface. They appear to rise from waters teeming with that minute surface-life which floors the ocean with organic remains.

The leading features of oceanic deposition must here be recalled to mind. The shallower parts of the ocean, to depths of about 2000 fathoms, accumulate sedimentary materials which never reach the deeper areas. The reason of this resides in the soluble nature of the calcareous tests of the dead foraminifera, coccoliths, &c., which contribute so largely to the sub-oceanic deposits. On these minute objects the waters act after the death of the tissues, slowly restoring the calcium carbonate to a state of solution. This action, inappreciable near the surface, increases in the depths with the increase of pressure. For a double reason, then, the test disappears before it reaches the greatest depths: the increasing solvent power of the water with its dissolved carbon dioxide, and the slower progress to the bottom of the dwindling particle, thereby ensuring longer exposure to the solvent actions. The endless shower of particles vanishes before it attains the greatest depths, much as a shower of snow in the upper atmosphere might dissolve to rain before the lower and warmer layers of the atmosphere were passed through. Hence it follows that on certain deep-lying parts of the ocean-floor the growth of sediment is extremely slow, and on other shallower parts the rate of growth is relatively fast.

Sir John Murray founded upon these facts a theory of coral-island formation which, unlike Darwin's, did not necessarily involve a subsiding sea-bottom. According to Murray's view (and A. Agassiz arrived at similar conclusions), the foundation of the island is a submarine bank (it may be a volcanic cone) which has been extended upwards by the accumulation of ooze, till the depth beneath the surface is so much reduced that the reef-builders find thereon a suitable habitat. Here the active agent is the sedimentary growth, and the coral polyp and lime-secreting algæ bring their edifices to the surface without any crustal movement being involved.

The claims of this ingenious alternative led to the boring of the coral island Funafuti by Sollas and David. The boring, which was carried to a depth of 185 fathoms by David, met with no foundational layer of chalk or calcareous ooze. This should appear at about 30 fathoms down if the foundation of the coral rock indeed rests upon ooze, and there had been no vertical movement. But as only true coral rock was encountered the entire 1100 feet of descent, clearly subsidence must be admitted. Although the true foundation was not reached by the boring, Darwin's view that there has been a subsidence is in this case established. The coral rock now at a depth below sea-level of 1100 feet must once have been at or near the surface.

The beautiful map of oceanic deposits, which science owes mainly to the work of Murray and Renard, shows that the areas on which the coral islands stand are, without exception, floored by the globigerina ooze, and, as already remarked, from 1000 to 2000 fathoms beneath the surface. Neighbouring areas sink to 3000 fathoms and more, and are covered with the red clay, or with radiolarian ooze. From these deep regions no coral islands spring. It is known that the growth of these two last-named deposits is of extreme slowness. In truth, the calcareous supplies being cut off, no materials beyond those furnished by silica-secreting organisms, and such volcanic dust as is carried through the atmosphere or through the water, and the meteoric dust from space, reach these remote parts of the world. Hence they become the charnel house of unburied remains; some, possibly, of even Tertiary age; others, the harder parts of the remains of recent cetaceans, reaching their present position probably at long-separated intervals of time. The strange manganese nodules also grow here, apparently from slowly decomposing mineral debris, and collecting considerable quantities of uranium. That the dredge often comes up laden with these unburied remains is proof of the paucity of sedimentary materials reaching these depths.

There is no subject in the domain of physical geology more fascinating than those refined but obscure laws of sedimentation which preserve to the continents their waste of minute detritus, so that they collect age after age in the shallow waters around the land, instead of being borne to the remote parts of the ocean. In these laws we recognize an important factor in mountain-building, although only a contributory one. For it appears certain that without their intervention all the finest debris from the continents would have been scattered far from the land, and the sediments of the great synclines of the earth proportionally reduced in volume. The amount so deducted might be one quarter of the whole, or even more. So great a loss must not only have resulted in a less mean elevation of the continents above sea-level (unless compensatory effects intervened), but also in lesser radiothermal effects in the sediments; delaying all the great events of mountain elevation and vulcanicity, which in their sum represent a large part of the dynamical history of the surface crust.

The experimental demonstration of these fundamental laws of sedimentation is easy. Measure out equal volumes of sea-water and distilled water in glass beakers. Stir into each an equal amount of finest kaolin or clay, and leave at rest. In a few hours, the sea-water is again limpid, while the fresh water for days and even weeks may keep its cloudiness. The dissolved salts are evidently responsible for the clearing, seeing that the conditions are alike in all other respects. Again, we may vary the experiment. We make up solutions of sodium chloride and of magnesium chloride in the same proportions as these salts exist in sea-water. We now find that the magnesium chloride clears more rapidly, although there is much less of this salt present than there is of chloride of sodium. And if we once more compare these two salts, but this time make up quantities proportional to the molecular weights of the salts, so that equal numbers of molecules are present in each solution, the difference is still more marked in favour of the magnesium chloride. Lastly we compare the sedimentation in a solution of sugar with that in distilled water, and find little or no difference.

The fundamental fact to be learned from these experiments, and which has been confirmed by scores of others of the most varied kind, is that the salt must be ionized in order to bring about the precipitation of the clay. Such solutions conduct electricity as the ionized air in the electroscope conveys a current, *i.e.* in virtue of the presence of free electric charges on certain of the atoms and dissociated molecules ; both positively and negatively charged ions being necessarily present in equal numbers or in equal electric quantity. In some way the free electric charge causes the suspended particles of clay to adhere one to another, and so they fall to the bottom ; the viscous opposition of the water being less effective on the aggregated particles. Sugar will not act, because sugar solution contains no ions. The magnesium is more effective than the sodium, because the divalent metallic ion of magnesium carries a double ionic charge, whereas the sodium ion is monovalent, and carries but the single ionic quantity of electricity.[1]

This effect, so important in geological history, and not without influence on the central oceanic deposits which we have been discussing, is, then, dependent on the fact that sea-water is rich in ions, divalent and monovalent. And these laws operate everywhere throughout the ocean. They hasten the finest particles, derived from broken-up organisms and atmospheric dust, on their journey to the bottom; they effectually secure the isolation of the central oceans from the land sediments; and, not improbably, they operate to secure the precipitation of ions of high valency, such as those of dissolved uranium, along with the falling sediments. The precipitation of the metallic ion has, indeed, in some cases been experimentally proved.[2] And this brings us back from our digression —if indeed it be a digression—to the radioactive condition of the sediments in the unstable oceanic areas.

From what has already been stated in a former chapter, when dealing with the radioactivity of the

[1] Whetham, *Phil. Mag.*, xlviii., p. 74; Joly, *Sci. Proc. Royal Dublin Society*, ix., p. 325, *Compte rendu Congrès Géol. Internat.*, 1900, p. 1 *et seq.*, and *Trans. R.D.S.*, vii., 391.

[2] Linder and Picton, *Chem. Soc. Journal*, 1895.

oceanic sediments, it would appear that the precipitation of the uranium occurs all over the ocean. Near the land it becomes entangled with the finer detrital sediments; further out it is associated with the chalky oozes derived from pelagic life; and in the deeper central seas it reaches the red clays. Its varying degree of concentration will then mainly depend on the abundance of non-radioactive materials falling along with it to the bottom, and acting as diluent. The slowly collecting red clays and radiolarian oozes attain a radioactivity from 30 to 50: the fast-forming globigerina ooze, from 6 to 8; and the near-shore deposits (in harmony with the finding on the land sediments), from 3 to 5.

Can we associate the instability of certain areas with the radioactivity of the sediments piled thereon? At first sight it might be thought that if this was so the instability should be most in evidence where the radioactivity was highest. But, in truth, the conditions requisite for the attainment of a high temperature at the base of a sedimentary deposit show that this inference is not justified. It was shown (p. 77) that the basal temperature was greater as the square of the depth of the deposit, so that a given amount of radium highly diluted with inert materials, and therefore contained in a deep layer of feeble radioactivity per unit of volume, will give rise to a higher temperature beneath than the same amount of radium concentrated in a shallow deposit, although the radioactivity is, of course, in the latter case greater. Now these are just the relative conditions prevailing in the globigerina ooze and in the red clay. The former has a lower radioactivity, but not necessarily a lesser total quantity of radium. It is highly dilute, and therefore the contained radium is more deeply buried. The latter with a greater radioactivity keeps its radium nearer the surface. The first distribution favours the retention of the heat; the second facilitates its escape.

Another condition also enters to enhance the thermal effects beneath the chalky deposit—the lesser conductivity of such deposits. We are dealing with what is essentially chalk. This is a substance of exceptionally low conductivity, the recorded determinations generally assigning 0·002 or 0·003 as the thermal conductivity: i.e., one-half or three-fourths of what may be commonly ascribed to sedimentary materials.

Thus there are present all the conditions favourable to a high basal temperature—a high radioactivity (6 to 8), a low conductivity, and a rapid rate of deposition. The question then naturally turns on the thickness of sediment which may be attained in this part of the ocean. The great subject of the deep-sea deposits is here presented to us with a new interest. Can we obtain no guidance as to what may be the thickness of these deposits over the disturbed areas of the ocean floor?

The time during which the accumulation may have been progressing seems indeterminate, and, for all we know, may have extended throughout the whole of geological history. This leads us to consider for a little the doctrine of the permanency of the oceans and continents.

It has been shown that the permanence of the great oceanic depressions may be involved in the larger dynamics of the earth;[1] and it is for future investigations to determine how far lost continents, of the magnitude which some geologists seem to favour, can, consistently with this important result, be supposed to come into existence and again disappear. But there is an independent line of evidence to show that continents comparable in magnitude with those we know, and of any comparable longevity, cannot have existed. The argument is based on the facts of solvent denudation.

The sodium in the ocean is the key to the history of solvent denudation. Every connected fact we know supports the view that this was derived from the igneous rocks of the earth by the processes of weathering and solution progressing throughout geological time. The total of it has been preserved, practically in its entirety, in the ocean, wherein the chloride has accumulated to such an extent that if spread over the earth's surface in a uniform layer, it would cover it to a depth of 112 feet. Contrasted with so great a mass, the quantities of rock-salt which have been abstracted from the sea by evaporation at various times in the past, are minute, and, as a source of error in our calculations, may be neglected.

Now it is found that the quantity of oceanic sodium tallies with the sediments as we find them upon the existing continents: that is, the estimated bulk of the sediments is just about what would account for the salinity of the ocean, assuming that these sediments

[1] Love, *Rep. British Assoc.*, 1907, and *Phil. Trans. R.S.*, 207, p. 171.

were derived from igneous rocks having the same average composition as those we now see upon the surface. The last assumption is so probable that it need not be further discussed. But if the known sediments are such as agree with the record of denudation which the ocean gives us, there cannot be yet other continents each with its own burden of sediments now hidden beneath the ocean. True, our estimate of sediment upon the existing land-surface may not be very accurate, but it can hardly be supposed miscalculated to the extent of the sedimentation of continents comparable in magnitude with those we are dealing with. The figures show that so far from there being any indication of missing sediment, it is the other way, and we are confronted with an *embarras de richesse* which accounts for the marine sodium, and leaves sedimentary material over.

Estimates of the amount of sediment on the earth have been made by several authorities. The early estimates were all very high. Dana thought they might be taken as not exceeding an average of 5 miles (8 kilometres) in depth over the continents. Mellard Reade modified an originally very high estimate to one of about the same magnitude as Dana's, stating his conclusion that the total deposits would cover the entire earth to a depth of one mile (1·61 kilometre). Van Hise,[1] reconsidering the estimates of these earlier writers, states, as a guess, that the sediments which exist in the zone of katamorphism, and have always

[1] Van Hise, *A Treatise on Metamorphism*, p. 939.

been in this zone (i.e., have not been thermally metamorphosed into different petrological materials) may be taken as on an average 2 kilometres over the continents. Prior to this estimate of Van Hise, the author had, for reasons which need not be cited, modified Mellard Reade's estimate to one which may be represented by saying that the purely detrital sediments would cover the continents to the depth of 1·1 mile (1·77 kilometre), including an allowance of 10 per cent. for coastal deposits. This estimate deals with those sediments with which alone we are now concerned. It will be seen that it is lower than any other available estimate, for the omitted limestones are not probably more than 5 per cent., and, besides, Van Hise's estimate is admittedly a restricted one.

A layer of sediment 1·77 kilometre thick spread over the area of the continents ($14{,}456 \times 10^4$ square kilometres) would possess a mass of about 64×10^{16} tonnes. And from the known facts of solvent weathering we are entitled to conclude that this mass is about 67 per cent. of the parent igneous rock from which it was derived. This gives to the parent rock a mass of $95{\cdot}5 \times 10^{16}$ tonnes.

The estimated amount of detrital sedimentaries must include about $8{\cdot}3 \times 10^{15}$ tonnes of the sodium oxide, Na_2O. This we conclude from the results of the analyses of typical sedimentary rocks given by Clarke in his *Data of Geochemistry*. The sodium in the ocean when converted into the equivalent mass of the oxide affords $21{\cdot}0 \times 10^{15}$ tonnes. It is evident that in the soda of the estimated sediments, added to the soda deduced from oceanic sodium, we have, apparently, the total soda which entered into the composition of the original igneous rock. We accordingly restore it, and find that the result is 3·07 per cent. of soda.

Let us now compare this result with the average soda content of igneous rocks. According to tables given by Clarke,[1] the average soda content of these rocks is 3·39 per cent. The agreement is close, considering the nature of the problem. It is noteworthy, however, that there is a *deficiency* of sodium. What should we infer from this? The only reasonable way of effecting a reconciliation is to reduce the amount of parent igneous rock, or, in other words, to reduce the estimate of the sediments. When we do so, we can of course bring about agreement (p. 37). Thus we find that the estimated sediments on the continents are *in excess* of the requirements of solvent denudation, as revealed by the oceanic record. How, then, can it be assumed that the sedimentary burden of whole continents is beneath the sea? Nor does it appear possible to assail the truth of the oceanic record, unless on the possible but improbable supposition that there was a large amount of original sodium in the ocean not derived from the rocks. But if we take this view, we limit still more the amount of sodium which we may restore to the igneous rocks, and so necessitate a further reduction in the estimate of the sediments.

The theory of lost continents has often been upheld in order to account for sedimentary accumulations in

[1] Clarke, *The Data of Geochemistry*, p. 31.

certain lands, which accumulations are, so to speak, fatherless. To postulate whole continents in order to supply the parentage is to get into the serious difficulty we have discussed above. That former continents of any antiquity or magnitude are not hidden beneath the waves seems certain, *unless the estimates of sediments are quite erroneous.* The argument, of course, does not touch on the former existence of those causeways which biological history requires as the highways of organic migration.

Our point here is, however, the essential antiquity of the great oceanic depressions which embrace the coral islands, and the greatness of the time during which sediments may have accumulated in those areas. It would appear that we may in no niggardly spirit deal with that reserve of time which geological science has laid apart for those who lawfully prefer their demands.

As regards the rate of deposition, it seems assured that this varies enormously with the nature of the deposit and the surface conditions, as well as with the conditions of depth. Even siliceous deposits may attain relatively rapid rates of growth where favoured by abundance of dissolved siliceous materials carried by the great rivers into the ocean.

Quite possibly this last condition has contributed to the amount of siliceous ooze which is presented to our inspection on the re-elevated island of Barbados; this island being not very far removed from the drainage discharge of a great continent. These deposits are, however, from their character, undoubtedly oceanic sediments, and the facts which they seem to convey are of particular interest. The history of the island as revealed in its sediments is remarkable. Underlying the deep-sea deposits are sandy clays and sandstones of relatively shallow-water character. Above the deep-sea beds are coral rocks of surface formation. The island must therefore have gone down to the depths for a period and re-emerged bearing with it the accumulations which gathered during the period of submergence. As it approached the surface, the reef-builder settled upon it, and covered the soft oceanic sediments with a protecting cover of coral. Thus we have in its rocks a fairly complete record of the recent sedimentary history of a small patch of the ocean floor.

In their investigation of the island Jukes-Browne and Harrison obtained measurements of the deposits which show that during the period of submergence a maximum of about 300 feet of sediments of oceanic type were laid down.[1] Their section on Mount Hillaby is as follows:—

Grey volcanic mudstones, ...	25 feet.
Very fine-grained argillaceous earths, red, pink, and yellowish, often mottled, .	60 ,,
Chalky earths and marls, ...	25 ,,
Calcareo-siliceous earths, with layers of pumiceous sand, ...	45 ,,
Pure siliceous radiolarian earths, . . .	130 ,,
Calcareo-siliceous earths, passing down into chalky earths, ...	40 ,,
Total, . . .	325 feet.

[1] Jukes-Browne and Harrison, *Q.J.G.S.* xlviii., p. 210.

The Scotland beds, of shallow-water origin, come immediately beneath. The writers show by other measurements that the real thickness is somewhat greater. Omitting the uppermost bed as being one of quite indeterminate rate of deposition, we have 300 feet of oceanic materials, mainly red clays (60 ft.) and radiolarian ooze (130 ft.); the other beds being seemingly an admixture of such sediments with some calcareous matter. Whatever the geological position of the Scotland beds may be, they are certainly not oceanic in origin; and again the rock immediately above is coral rock, and therefore marks the re-ascent of the submerged area to the surface.

We could, of course, form some idea of the rate of deposition of these oceanic beds if we knew the duration of submergence. Here we have to face conflicting views as to the precise geological period in which the submergence took place. Jukes-Browne and Harrison believe the beds collected probably during Pliocene and partly, perhaps, in Pleistocene times. J. W. Spencer[1] in a more recent paper ascribes these deposits to Eocene times. Sollas[2] estimates the maximum thickness of Eocene sediments as 12,000 feet. The rate of deposition if taken at one foot in a century would ascribe 1·2 million of years for the duration of this period. This rate is probably too great. One third of a foot as the mean rate of detrital sedimentation seems to accord better with the evidence as to the duration of geological time and the estimated

[1] Spencer, *Q.J.G.S.* lviii., p. 304 *et seq.*

[2] Sollas, *The Age of the Earth*, p. 14.

total maximum sedimentation of all the geological periods. On this rate the Eocene period must have lasted 3·6 millions of years. The whole of the Pliocene, following the same mode of estimation, would amount to 1·5 million years, and the Pliocene, Pleistocene, and Recent together to 2·7 millions of years. We see, then, that the period of submergence should be something between one million and three million years. These limits would give rates of deposition of the Barbados sediments from 9 to 3 millimetres in a century.

These rates appear very slow, but in point of fact they are too high to be generally assumed for such deposits. We see this when we reckon that if geological time extended 100 million years, a rate of 5 millimetres in a century would accumulate to 5 kilometres—a much greater depth than is allowable on our former estimate of the total amount of oceanic deposits.

It is, however, quite certain, as already remarked, that the rate of accumulation of oceanic sediments varies enormously. The chalk attained a thickness of from 1200 to 1500 ft. in British Cretaceous seas; and this is a true oceanic deposit. The period of deposition was a relatively brief one. Huxley inferred that the rate of collection of this ooze was as much as 1 inch in a year; but Sollas has shown that this rate is certainly excessive, and believes that 1 inch in 40 years would be more nearly correct.[1] Sir John Murray and Mr. Peake[2] have given reasons for supposing that in parts of the

[1] Sollas, *The Age of the Earth*, p. 160.

[2] Sir John Murray and R. E. Peake, *Royal Geographical Society Suppl. Papers*, 1904, p. 21.

Atlantic the Globigerina ooze collects on the cables at a rate of as much as 10 inches in a century. These estimates are in no case, of course, assured to us as correct; but it is apparent that the rate of accumulation of the calcareous ooze must be very much greater than that of the red clay, which cannot be more than a millimetre or two in a century. This lack of knowledge will certainly be remedied with the progress of science. There should be no insurmountable difficulty either as regards natural obstacles or costliness, in effecting direct measurements of the rate of collection of the faster-growing deposits.

We are, at the present time, left with only this much guidance as regards the rate of sedimentation progressing over the coral island plateaus, that it may be very considerable, and certainly *is* very considerable if at all resembling in rate that with which the chalk was laid down in the Cretaceous seas. Thus there is nothing improbable in the view that great accumulations of sediment exist around the coral islands. We appear restricted neither in our demands upon time nor as to rate of deposition—of course within limits. And we are assured that the radioactivity is more favourable for the attainment of larger radio-thermal effects than in ordinary sediments. If the rate of collecting had been only 5 millimetres in a century over a period considerably short of the probable duration of geological time, the depth of sediment might well amount to 4 kilometres, which, on the assumptions discussed in Chapter V, involves a reduction of effective strength amounting to nearly 10 per cent. But as the radioactivity of the chalky deposit is more than double that assumed for the sediments in preparing the table, and, again, as the conductivity is less for the sedimentary part of the total layer, the result tabulated must fall considerably short of the actual effect in the present case. Probably not less than 20 per cent. should be taken as the reduction in effective resistance and thickness of the crust due to an accumulation such as is here supposed. And, without improbable demands on the limiting total of the suboceanic sediments, the area of weakness might be sufficiently extended to determine a wide-spread upheaval of the central Pacific floor.

We have now in outline reviewed the conditions, both those which are known to prevail, and those which may reasonably be supposed to prevail, in the coral seas; and if we are right, the deeper sedimentary accumulations of the central ocean must play an important part in the movements of the crust. But we cannot stop here: the consequences of these movements must be considered, and, such as they are, they must be accredited to the influence of uranium when its energy is focussed, as it were, upon particular areas of the stressed and precarious crust of the earth. It may now turn out that those transgressions of the seas upon the continents, which in the past were of world-wide extent, in so far that they seem to have simultaneously submerged many lands, are traceable to these oceanic movements, or at least partially traceable thereto. If the buckling of the sea-floor may be defined by the estimates which Darwin and others have made of the recent settlement indicated by the coral islands, then this view seems more than mere hypothesis, for the rise and fall of the sea upon the land by some hundreds of feet is assured even if the continents be supposed to have remained rigidly resistant and fixed in their present position under the increased load.

The boring of Funafuti has shown that this island has subsided at least 900 or 1000 feet in recent times. For this the evidence seems quite irrefragable. But the boring which proves this extends but a small fraction of the probable depth of the coral. The penetration of the drill has hardly been more than a tenth or a twelfth part of the elevation of the island above the sea-floor. That the coral ends just beneath the boring is unlikely. We may fairly conclude that a much greater downward movement than 1000 feet has actually occurred. Darwin concluded from the evidence of the soundings that the whole subsidence cannot be considered less than 6000 feet, and believed that it began in Tertiary times. There is, in the case of the quasi-coral island of Barbados, evidence of a subsidence and re-elevation in Tertiary times of not less than 10,000 feet. Jukes-Browne and Harrison believe that in its case we may safely assume between 12,000 and 18,000 feet.[1]

Now, if we accept Darwin's estimate of 1000 fathoms for the general subsidence of the floor of the Coral Sea, the change in the capacity of the Pacific basin would have affected the distribution of land and water very considerably. It is easy to show, similarly, that if there

[1] Jukes-Browne and Harrison, *loc. cit.*, p. 201.

was an elevation of this amount from the present level, extensive areas of the existing continents would be covered. Taking the upheaved area at 2×10^7 square kilometres, the general level of the ocean would be raised about 340 feet, assuming a transgression of the waters to this depth upon five per cent. of the land area. The actual area of submerged land would be greater, but the average depth of submergence less than 340 feet. This depth would suffice for the deposition of the chalk.[1]

We do not here seek to account for the origin of the surface stress which is responsible for these movements, nor yet to account for such compensatory motions of the crust as probably occurred. We know such movements have taken place in past times; and whether they be referable to secular contraction or to adjustments of the lithosphere such as Love has suggested,[2] or to the operation of both causes, is outside the scope of our present subject.

The whole matter is a relatively small one, although to the earth's inhabitants so important. Imagine a globe 2 metres in diameter: the movement of the sea-floor we have assumed would be represented on such a globe by the radial displacement of an area upon its surface through about the one-fourth part of a millimetre, that is, about the one-hundredth part of an inch, or, say, the thickness of a visiting-card! The move-

[1] Jeffreys, Presidential Address, British Association, 1877; also Chamberlin and Salisbury, *Geology*, vol. iii., p. 150. In the above estimate the existing area of the ocean is taken as $36,708 \times 10^4$, and the present land surface as $14,456 \times 10^4$ square kilometres.

[2] Love, *loc. cit.*, p. 239 *et seq.*

ments in the geosynclines and the buckling of the ocean-floor are, indeed, trifling details of the earth's vicissitudes.

Those who take pleasure in the contemplation of the means—various in detail, alike in result—by which the ends of nature are often accomplished, will reflect on the fact that the foregoing views have led us to ascribe two dynamical episodes of the crust—fateful to organic history if petty in themselves—to the convective movements of uranium on the surface of the earth: the one restricted to the continental margins, owing to the solid nature of the materials involved; the other coming into existence in the remotest parts of the ocean-floor, the dissolved elements receding indefinitely from the land. In each case the convective and cumulative conditions have to be fulfilled; but the physical processes responsible are widely different.

CHAPTER VII.

RADIOACTIVITY AND MOUNTAIN STRUCTURE.

PROBABLE PERIODS OF TIME INVOLVED IN THE GENESIS OF THE ALPS —EVIDENCE OF HIGH TEMPERATURE IN ALPINE ROCKS—THESE CONDITIONS UNACCOUNTABLE WITHOUT THE INTERVENTION OF RADIOTHERMAL ACTIONS—RECENT VIEWS ON ALPINE TECTONICS—THE ORIGIN OF THE PRÉALPES AND OF THE GREAT LIMESTONE ALPS—THE DÉFERLEMENT OF THE RECUMBENT FOLDS—TEMPERATURES PROBABLY ATTAINED—THEORY OF THE HORST UNTENABLE —THE FLUID OR VISCOUS CONDITIONS SUFFICIENT TO ACCOUNT FOR THE RECUMBENCE OF THE FOLDS AND FOR THE DÉFERLEMENT—SKETCH OF THE STAGES OF DEVELOPMENT.

An intelligent observer of the geological changes progressing in southern Europe in Eocene times would have seen little to inspire him with a premonition of the events then developing. The Nummulitic limestones were being laid down in that enlarged Mediterranean which at this period, save for a few islands, covered most of south Europe. Of these stratified remains, as well as of the great beds of Cretaceous, Jurassic, Triassic, and Permian sediments beneath, our hypothetical observer would probably have been regardless; just as to-day we observe, with an indifference born of our transitoriness, the deposits rapidly gathering wherever river discharge is distributing the sediments over the sea-floor, or the lime-secreting organisms are actively at work. And yet it took but a few millions of years to uplift the deposits of the ancient Tethys;

pile high its sediments in fold upon fold in the Alps, the Carpathians, and the Himalayas; and—exposing them to the rigours of denudation at altitudes where glaciation, land-slip, and torrent for ever prevail—inaugurate a new epoch of sedimentation and upheaval.

In the case of the Alps, to which we wish now specially to refer, the chief upheaval appears to have been in Oligocene times, although movement continued to the close of the Pliocene. There was thus a period of some millions of years within which the entire phenomena were comprised. Again availing ourselves of Professor Sollas' computations, we may sum the maximum depths of sedimentary deposits of the geological periods concerned as follows:—

Pliocene, . .	5,000 feet,
Miocene, . .	9,000 ,,
Oligocene, . .	12,000 ,,
Eocene, . .	12,000 ,,

and assuming that the orogenic forces began their work in the last quarter of the Eocene period, we have a total of 29,000 feet as some measure of the time which elapsed. At the rate of one-third of a foot in a century these deposits could not have collected in less than 8·7 millions of years. It would appear that not less than from 6 to 8 millions of years were consumed in the genesis of the Alps before constructive movements finally ceased.

The progress of the earth-movements was attended by the usual volcanic phenomena. The Oligocene and Miocene volcanoes extended in a band marked by the Auvergne, the Eifel, the Bohemian, and the eastern

Carpathian eruptions; and, later, towards the close of the movements in Pliocene times, the south border regions of the Alps became the scene of eruptions such as those of Etna, Santorin, Somma (Vesuvius), &c.

We have referred to these well-known episodes with two objects in view: to recall to mind the time-interval involved, and the evidence of intense crustal disturbance, both dynamic and thermal. According to views explained in a previous chapter, the energetic effects of radium in the sediments and upper crust were a principal factor in localizing and bringing about these results. We propose now to inquire if, also, in the more intimate structure of the Alps, the radioactive energy may not have borne a part.

What we see to-day in the Alps is but a residue spared by denudation. It is certain that vast thicknesses of material have disappeared. Even while constructive effects were still in progress, denudative forces were not idle. Of this fact the shingle accumulations of the Molasse, where, on the northern borders of the Alps, they stand piled into mountains, bear eloquent testimony. In the sub-Apennine series of Italy, the great beds of clays, marls, and limestones afford evidence of these destructive processes continued into Pliocene times. We have already referred to Schmidt's estimate that the sedimentary covering must have in places amounted to from 15,000 to 20,000 metres. The evidence for this is mainly tectonic or structural; but is partly forthcoming in the changes which the materials now open to our inspection plainly reveal. Thus it is impossible to suppose that gneissic rocks can become so far plastic as to flow in and around the calcareous sediments, or be penetrated by the latter—as we see in the Jungfrau and elsewhere—unless great pressures and high temperatures prevailed. And, according to some writers, the temperatures revealed by the intimate structural changes of rock-forming minerals must have amounted to those of fusion. The existence of such conditions is supported by the observation that where the crystallization is now the most perfect, the phenomena of folding and injection are best developed.[1] These high temperatures would appear to be unaccountable without the intervention of radiothermal effects; and, indeed, have been regarded as enigmatic by observers of the phenomena in question. A covering of 20,000 metres in thickness would not occasion an earth-temperature exceeding 500° C. if the gradients were such as obtain in mountain regions generally; and 600° is about the limit we could ascribe to the purely passive effects of such a layer in elevating the geotherms.

Those who are still unacquainted with the recently published observations on the structure of the Alps may find it difficult to enter into what has now to be stated; for the facts are, indeed, very different from the generally preconceived ideas of mountain formation. Nor can we wonder that many geologists for long held back from admitting views which appeared so extreme. Receptivity is the first virtue of the scientific mind; but, with every desire to lay aside prejudice, many felt unequal to the acceptance of structural features

[1] Weinschenk, *C. R. Congrès Géol.*, 1900, p. 321 *et seq.*

involving a folding of the earth-crust in laps which lay for scores of miles from country to country, and the carriage of mountainous materials from the south of the Alps to the north, leaving them finally as Alpine ranges of ancient sediments reposing on foundations of more recent date. The historian of the subject will have to relate how some who finally were most active in advancing the new views were at first opposed to them. In the change of conviction of these eminent geologists we have the strongest proof of the convincing nature of the observations and the reality of the tectonic features upon which the recent views are founded.

The lesser mountains which stand along the northern border of the great limestone Alps, those known as the Préalpes, present the strange characteristic of resting upon materials younger than themselves. Such mountains as the remarkable-looking Mythen, near Schwyz, for instance, are weathered from masses of Triassic and Jurassic rock, and repose on the much more recent Flysch. In sharp contrast to the Flysch scenery, they stand as abrupt and gigantic erratics, which have been transported from the central zone of the Alps lying far to the south. They are strangers petrologically, stratographically, and geographically,[1] to the locality in which they now occur. The exotic materials may be dolomites, limestones, schists, sandstones, or rocks of igneous origin. They show in every case traces of the severe dynamic actions to which they have been subjected in transit. The igneous, like the sedimentary,

[1] De Lapparent, *Traité de Géologie*, p. 1785.

klippen, can be traced to distant sources; to the massif of Belladonne, to Mont Blanc, Lugano, and the Tyrol. The Préalpes are, in fact, mountains without local roots.

In this last-named essential feature, the Préalpes do not differ from the still greater limestone Alps which succeed them to the south. These giants, e.g., the Jungfrau, Wetterhorn, Eiger, &c., are also without local foundations. They have been formed from the overthrown and drawn-out anticlines of great crust-folds, whose synclines or roots are traceable to the south side of the Rhone Valley. The Bernese Oberland originated in the piling-up of four great sheets or recumbent folds, one of which is continued into the Préalpes. With Lugeon[1] we may see in the phenomenon of the formation of the Préalpes a detail; regarding it as a normal expression of that mechanism which has created the Swiss Alps. For these limestone masses of the Oberland are not indications of a merely local shift of the sedimentary covering of the Alps. Almost the whole covering has been pushed over and piled up to the north. Lugeon[2] concludes that, before denudation had done its work and cut off the Préalpes from their roots, there would have been found sheets to the number of eight superimposed and extending between the Mont Blanc massif and the massif of the Finsteraarhorn: these sheets being the overthrown folds of the wrinkled sedimentary covering. The general nature of the alpine structure will be

[1] Lugeon, *Bulletin Soc. Géol. de France*, 1901, p. 772.

[2] Lugeon, *loc. cit.*

understood from the presentation of it diagrammatically after Schmidt of Basel (Plate V.).[1] The section extends from north to south, and brings out the relations of the several recumbent folds. We must imagine almost the whole of these superimposed folds now removed from the central regions of the Alps by denudation, and leaving the underlying gneisses rising through the remains of Permian, Triassic, and Jurassic sediments; while to the north the great limestone mountains, and further north still, the Préalpes, carvèd from the remains of the recumbent folds, now stand with almost as little resemblance to the vanished mountains as the memories of the past have to its intense reality.

These views as to the origin of the Alps, which are shared at the present day by so many distinguished geologists, had their origin in the labours of many now gone; dating back to Studer; finding their inspiration in the work of Heim, Suess, and Marcel Bertrand; and their consummation in that of Lugeon, Schardt, Rothpletz, Schmidt, and many others. Nor must it be forgotten that nearer home, somewhat similar phenomena, necessarily on a smaller scale, were recognized by Lapworth, twenty-six years ago, in his work on the structure of the Scottish Highlands.

An important tectonic principle underlies the development of the phenomena we have just been reviewing. The uppermost of the superimposed recumbent folds is more extended in its development than those which lie beneath. Passing downwards from the

[1] Schmidt, *Ec. Geol. Helvetiae*, vol. ix., No. 4.

PLATE V.

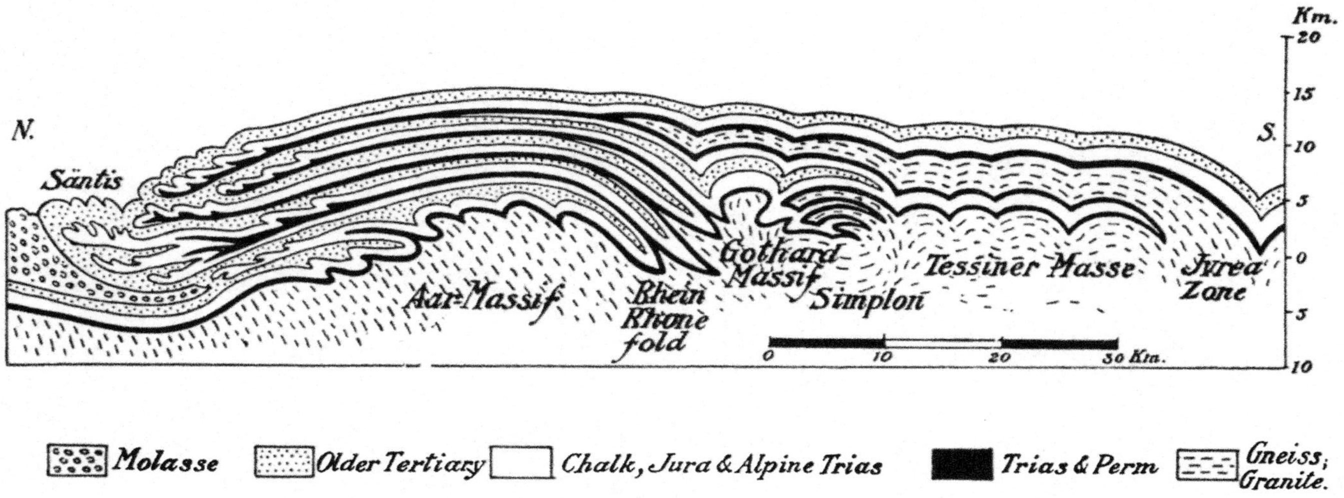

DIAGRAMMATIC SECTION THROUGH THE SWISS ALPS (AFTER SCHMIDT).

highest of the folds, they are found to be less and less extended both in the northerly and in the southerly direction, speaking of the special case—the Alps—now before us. This feature might be described somewhat differently. We might say that those folds which had their roots further to the south were the most drawn-out towards the north: or again we might say that the synclinal or deep-seated part of the fold has lagged behind the anticlinal or what was originally the highest part of the fold, in the advance of the latter to the north. The anticline has advanced relatively to the syncline. To this law one exception only is observed in the Swiss Alps; the sheet of the Brèche (*breccien decke*) falls short, in its northerly extension, of the underlying fold, which extends to form the Préalpes.

Contemplating such a generalized section as Professor Schmidt's, or, indeed, more particular sections, such as those in the Mont Blanc Massif by Marcel Bertrand,[1] of the Dent de Morcles, Diablerets, Wildhorn, and Massif de la Brèche by Lugeon,[2] or finally Termier's section of the Pelvoux Massif,[3] one is reminded of the breaking of waves on a sloping beach. The wave, retarded at its base, is carried forward above by its inertia, and finally spreads far up on the strand; and if it could there remain, the succeeding wave must necessarily find itself superimposed upon the first. But no effects of inertia, no kinetic effects, may be called to our

[1] Marcel Bertrand, *Cong. Géol. Internat.*, 1900, *Guide Géol.*, xiii. a, p. 44. [2] Lugeon, *loc. cit.*, p. 773.

[3] De Lapparent, *Traité de Géol.*, p. 1773.

aid in explaining the formation of mountains. Some geologists have accordingly supposed that in order to account for the recumbent folds and the peculiar phenomena of increasing overlap, or *déferlement*, an obstacle, fixed and deep-seated, must have arrested the roots or synclines of the folds, and held them against translational motion, while a movement of the upper crust drew out and carried forward the anticlines. Others have contented themselves by recording the facts without advancing any explanatory hypothesis beyond that embodied in the incontestable statement that such phenomena must be referred to the effects of tangential forces acting in the earth's crust.

It would appear that the explanation of the phenomena of recumbent folds and their *déferlement* is to be obtained directly from the temperature conditions prevailing throughout the stressed pile of rocks; and here the subject of mountain tectonics touches that with which we are specially concerned—the geological influence of accumulated radioactive energy.

As already remarked (p. 110), a rise of temperature due to this source of from 300° to 500° might be added to such temperatures as would arise from the mere blanketing of the earth, and the consequent upward movement of the geotherms. The time element is here the most important consideration. The whole sequence of events from the first orogenic movements to the final upheaval in Pliocene times must probably have been comprised within a period of about six to eight million years, although we cannot be sure that a longer time was not involved

Unfortunately the full investigation of the distribution of temperature after any given time is beset with difficulties; the conditions being extremely complex. If the radioactive heating was strictly adiabatic—that is, if all the heat was conserved and none entered from without—the time required for the attainment of the equilibrium radioactive temperature of 400° would be just about six million years, assuming the lower estimates of the radioactivity of the Simplon rocks as indicative of the radioactive state of the materials. The conditions are not, indeed, adiabatic; but, on the other hand, the rocks upraised by lateral pressure were by no means at 0°C. to start with. They must be assumed to have possessed such temperatures as the prior radiothermal effects, and the conducted heat from the earth's interior, may have established.

It would from this appear probable that if durations of even five or six million years were involved, the equilibrium radioactive temperatures must nearly have been attained. The effects of heat conducted from the underlying earth-crust have to be added (and here again radiothermal effects must be taken into account, according to the principles explained in Chapter V), leading to a further rise in temperature of not less than 500° or 600°. In such considerations the observed indications of high temperatures in materials now laid bare by denudation, probably find their explanation.

The first fact that we infer from the former existence of such a temperature distribution is the improbability, indeed the impossibility, that anything resembling a rigid obstacle, or deep-seated horst, can have existed beneath the present surface-level, and opposed the northerly movement of the deep-lying synclines. For such a horst can only have been constituted of some siliceous rock-material such as we find everywhere rising through the worn-down sediments of the Alps; and the idea that this could retain rigidity under the prevailing temperature conditions, must be dismissed. There is no need to labour this question; the horst cannot have existed. To what, then, is the retardation of the lower parts of the folds, their overthrow, above, to the north, and their *déferlement*, to be ascribed?

A little consideration shows that the very conditions of high temperature and viscosity, which render untenable the hypothesis of a rigid obstacle, suffice to afford a full explanation of the retardation of the roots of the folds. For directed translatory movements cannot be transmitted through a fluid, the pressure in which is necessarily hydrostatic, and must be exerted equally in every direction. And this applies, not only to a fluid, but to a body which will yield viscously to an impressed force. There will be a gradation, according as viscosity gives place to rigidity, between the states in which the applied force resolves itself into a purely hydrostatic pressure, and in which it is transmitted through the material as a directed thrust. The nature of the force, in the most general case, of course, has to be considered; whether it is suddenly applied and of brief duration, or steady and long-continued. The latter conditions alone apply to the present case.

It follows from this that, although a tangential force

or pressure be engendered by a crustal movement occurring to the south, and the resultant effects be transmitted northwards, these stresses can only mechanically affect the rigid parts of the crust into which they are carried. That is to say, they may result in folding and crushing, or horizontally transporting, the upper layers of the earth's crust; but in the deeper-lying viscous materials they must be resolved into hydrostatic pressure which may act to upheave the overlying covering, but must refuse to transmit the horizontal translatory movements affecting the rigid materials above.

Between the regions in which these two opposing conditions prevail there will be no hard and fast line; but with the downward increase of viscosity there will be a gradual failure of the mechanical conditions and an increase of the hydrostatic. Thus while the uppermost layers of the crust may be transported to the full amount of the crustal displacement acting from the south (speaking still of the Alps), deeper down there will be a lesser horizontal movement, and still deeper there is no influence to urge the viscous rock-materials in a northerly direction. The consequences of these conditions must be the recumbence of the folds formed under the crust-stress, and their *déferlement* towards the north. To see this, we must follow the several stages of development.

The earliest movements, we may suppose, result in flexures of the Jura-Mountain type—that is, in a succession of undulations more or less symmetrical. As the orogenic force continues and develops, these undulations give place to folds, the limbs of which are approximately vertical, and the synclinal parts of which become ever more and more depressed into the deeper, and necessarily hotter, underlying materials; the anticlines being probably correspondingly elevated. These events are slowly developed, and the temperature beneath is steadily rising in consequence of the conducted interior heat, and the steady accumulation of radioactive energy in the sedimentary rocks and in the buried radioactive layer of the earth. The work expended on the crushed and sheared rock also contributes to the developing temperature. Thus the geotherms must move upwards, and the viscous conditions extend from below; continually diminishing the downward range of the translatory movements progressing in the higher parts. While above the folded sediments are being carried northward, beneath they are becoming anchored in the growing viscosity of the medium. The anticlines will bend over, and the most southerly of the folds will gradually become pushed or bent over those lying to the north. Finally, the whole upper part of the sheaf will become horizontally recumbent; and as the uppermost folds will be those experiencing the greatest effects of the continued displacement, the *déferlement* or overlap must necessarily arise.

We may follow these stages of mountain evolution in a diagram (Plate VI) in which we eliminate intermediate conditions, and regard the early and final stages of development only. In the upper sketch we suppose the lateral compression much developed and the upward movement of the geotherms in progress. The

PLATE VI.

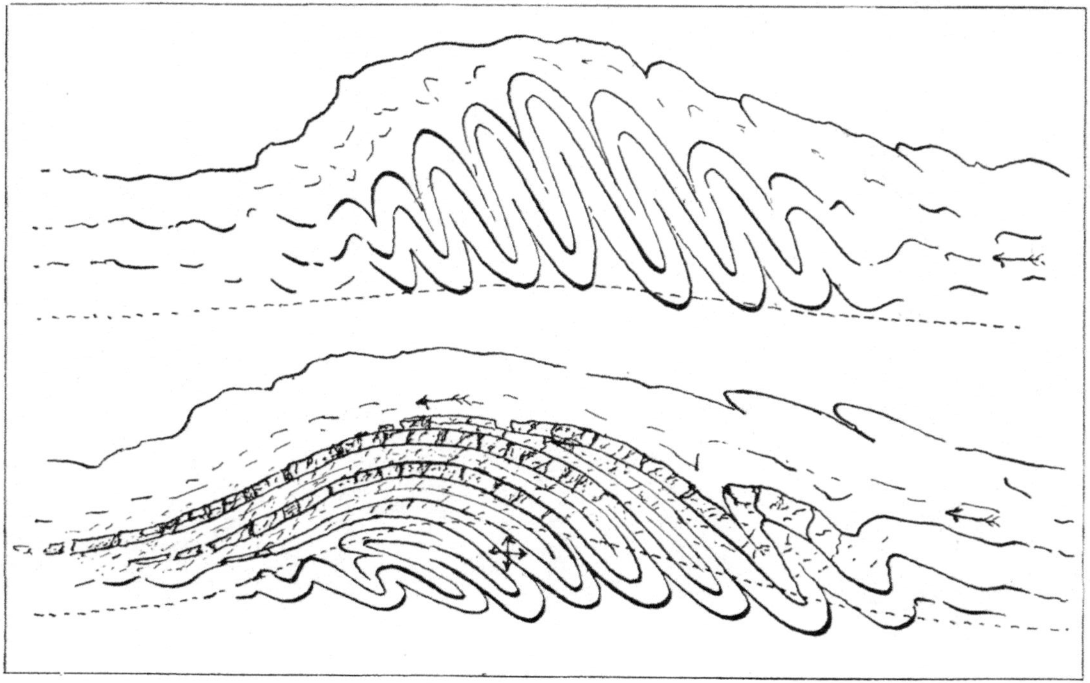

FORMATION OF RECUMBENT FOLDS.

dotted line may be assumed to be a geotherm having a temperature of viscosity. If the conditions here shown persist indefinitely, there is no doubt that the only further developments possible are the continued crushing of the sediments and the bodily displacement of the whole mass to the north. The second figure is intended to show in what manner these results are evaded. The geotherm of viscosity has risen. All above it is affected mechanically by the continuing stress, and borne northwards in varying degree depending upon the rigidity. The folds have been overthrown and drawn out; those which lay originally most to the south have become the uppermost; and, experiencing the maximum amount of displacement, overlap those lying beneath. There has also been a certain amount of upthrow owing to the hydrostatic pressure. This last-mentioned element of the phenomena is of highly indeterminate character, for we know not the limits to which the hydrostatic pressure may be transmitted, and where it may most readily find relief. While, according to some of the published sections, it would seem to have influenced the final results of the orogenic movements, a discussion of its consequences would not be profitable.

CHAPTER VIII.

RADIOACTIVITY AND THE INTERIOR OF THE EARTH.

DIFFICULTIES INVOLVED IN ASSUMING THE EARTH TO BE MERELY A COOLING BODY—KELVIN'S CALCULATION OF THE AGE OF THE EARTH AT VARIANCE WITH THE INDICATIONS OF OTHER METHODS—FROM THIS AN ARGUMENT FOR THE RADIOACTIVE HEATING OF THE EARTH—THE FURTHER ARGUMENT FROM THE OBSERVED SUFFICIENCY OF THE RADIOACTIVE SUBSTANCES IN THE CRUST AND FROM THE COMMON DESTINY OF ALL RADIOACTIVE HEAVENLY BODIES—DIFFICULTIES ATTENDING THE INVESTIGATION OF THE INTERIOR URANIUM—KELVIN'S PROOF OF THE THERMAL ISOLATION OF THE INTERIOR—THIS REMOVES THE DIFFICULTY OF ASSUMING THE EXISTENCE OF RADIOACTIVITY IN THE CENTRAL PARTS OF THE EARTH—ARGUMENTS IN FAVOUR OF SUPPOSING A WIDESPREAD DISTRIBUTION OF URANIUM IN THE EARTH—THE ULTIMATE CONSEQUENCES OF INTERIOR RADIOACTIVITY—POSSIBILITIES OF THE PAST AND OF THE FUTURE.

THE fact that heat is continually passing out at the surface of the earth is conclusive evidence of some source of heat within. Until the discovery of the heat-producing properties of radium it was assumed without controversy, it may be said, that the source of heat was the original heat with which the world was endowed at the time of its aggregation, and which was probably ultimately of dynamical origin. There was, indeed, strong evidence to support this view. That the earth had once been at a higher temperature was to be inferred directly, not only from an examination of the crust, but by the mere logical process of reversing the heat-losses of past time. Again, the facts of mountain elevation to which we have referred required a cooling and shrinking magma as the precursor of the existing crust. It was but natural to draw the inference that the energy giving rise to such conditions, still lingering in the materials of the earth, continued to escape at the surface.

The theory of the former high temperature of the earth appears well founded on observation; but when it is sought to enter more particularly into the thermal history of the earth, difficulties present themselves. These difficulties first became apparent in the work which Lord Kelvin directed towards determining the time which had elapsed since the earth's cooling from a temperature of solidification.[1]

The principles underlying this work are easily grasped. Kelvin started with a world at 7000° F. (3710° C.), at which temperature he assumed the cessation of convective currents, with the freezing of the materials of which it is composed. We have, then, a sphere 8000 miles in diameter at a uniform temperature of 3710°, and cooling from the surface inwards due to the continued loss by radiation. Assuming the actual surface to stand always at a temperature approximating to its present surface temperature, and so getting rid of a radiation coefficient, the problem is to calculate how long a period must elapse before the temperature has fallen from the surface inwards so as to afford the average thermal gradient which to-day prevails in the outer parts. The solution of this problem, of course, involves the conductivity of the rocks or other materials through which the thermal flux to the surface takes place.

[1] Kelvin, *Trans. Royal Soc. Edin.*, 1862, vol. xxiii.

Lord Kelvin's earlier calculations led to periods of from 20 to 400 millions of years; but later, as the constants entering into his work became replaced by more reliable data, his final conclusions pointed to from 20 to 40 millions of years, and more nearly 20 millions than 40 millions.[1]

But this conclusion is at variance with other methods of determining the age of the earth: for other methods, based upon the accumulated effects of certain processes going on at the surface, exist. These geological methods rely, in short, on surface changes which have taken place since the waters of the ocean condensed and rain and rivers began their work. The period of time included in the "geological age" must therefore be somewhat less—not very much possibly—than what is included in the period estimated by Kelvin: that is, from the *consistentior status*, or state of terrestrial interior stability. Nevertheless, the geological age, no matter how we approach its estimation, consistently yields a greater interval than Kelvin's 20 to 40 millions of years.

The method of finding the geological age by the accumulation of the sediments owes much to Sir Archibald Geikie. In it the total column of the sedimentary rocks, as derived from the integration of all geological strata where measured at their maxima, serves as numerator, and the denominator is sought in the rate at which sediment is borne from the land and collected on the areas of deposition. Geikie has deduced an age of about 100 million years on this basis.[2]

[1] Kelvin, *Phil. Mag.* 47, p. 66, 1889.

[2] Geikie, *Presidential Address to Sect. C., Brit. Assoc. Rep.*, 1899.

There is, indeed, great difficulty in arriving at the required data; and it cannot be said that up to the present the final verdict deducible from this method has been heard. Sollas, assuming a high rate of deposition, arrives by this method at a lesser age—26 millions of years.[1] It will be seen in a later chapter that the sources of error attending this method would most probably lead to an underestimate of geological time.

A second method is to proceed on the facts, more accessible, of solvent denudation, and to take the accumulation of dissolved substances in the ocean as numerator. Stated, however, in these terms the method will not work; for, with one exception, the substances brought by the rivers to the ocean are again abstracted from it. Results obtained by using the method without attention to this point are valueless. However, in the element sodium a highly soluble substance is found, which is not abstracted unless by accidental impounding and evaporation of ocean lagoons, &c.—but these losses are negligible.[2] As the ascertainment of the total sodium in the sea is easily accomplished without serious error, it only remains to find the denominator in the annual river-supply of the element sodium. Available data respecting the quantity of this element in average river-water, and the amount of the latter annually reaching the oceans, give a value of about 90 to 100 millions of years as the result obtained by this method. There is no doubt that before long more reliable values of the denominator will be forthcoming.

[1] Sollas, *The Age of the Earth*, p. 37 *et seq.*

[2] Joly, *Trans. Royal Dublin Soc.*, vol. vii., ser. ii., p. 23 *et seq.*

Both these geological methods proceed on the assumption that there has been sensible uniformity in geological processes throughout the period it is sought to determine. The arguments in favour of this assumption cannot be enlarged upon here. We cannot suppose these methods to err threefold or fourfold on the side of excess. Latterly the method of fixing the ages of the geological strata by the helium accumulated in radioactive minerals, as used by Rutherford, and recently systematically applied by Strutt, point to ages not less, but possibly even greater, than the geological methods indicate. The subject is more fully dealt with in Chapter XI.

We have then an evident contradiction between the physical method founded on the cooling of the earth and other methods. This suggests that it may be erroneous to assume that the earth is merely a cooling body, losing from a primitive store of heat, like an ember dropped from a furnace. No explanation of the discrepancy was forthcoming till the principles of radioactive heating were made known, and the fact was established that radium was a widely diffused, if much-diluted, surface constituent of the crust. It was then apparent that the difficulty could be bridged by supposing that the heat-loss of the earth had during a part of geological time been made good by radioactive supplies.[1] The work of Strutt finally showed that there was actually an *embarras de richesse* of radium in the surface rocks.

We have then, so far, found two good reasons for

[1] Joly, *Nature*, October 1, 1903.

accepting the view that the earth's escaping heat is mainly radiothermal in origin: the fact that this view reconciles the physical and geological methods of determining the age of the earth; and the fact that *the radium is there.*

Let us amplify this last reason by a homely illustration. We find ourselves in a room in which we see a bright and ample fire. Are we justified in ascribing the temperature of the apartment to the visible source of heat? There may be other sources of heat; indeed we may argue, and argue rightly, that the fire is not responsible for the more primitive heat of the walls and the building; but are we justified in assuming that the heat which is continually escaping is supplied by the fire? The analogy is really exact, for if, on the one hand, it be urged that we know not the depth to which the surface radioactivity of the earth extends, we may also say—reverting to the illustration—we are without knowledge of the actual heat-flow from the fire. Only in one particular is the analogy unfaithful. We know the radiothermal source is all but eternal in duration, and therefore we possess an assurance of sufficiency which does not exist in the case of the illustration.

But we find another reason for the belief that the earth is now, and has for long been, losing radioactive heat, and only cooling so fast as the radioactive materials themselves decay, in the broad generalization that in accepting this view we hardly do more than anticipate an inevitable condition.

Every solar and every planetary body originating as

a dynamically heated aggregate, and containing above certain small amount of uranium among its constituents, must pass finally into the stage of the radiothermal sustentation of its temperature, unless the original heat of the body and its capacity of retaining it are so preponderating that they outlive the radioactive elements. For in the beginning the heat-loss by radiation is great, and so the primitive capital is run down, and with its diminution and the cooling of the surface the rate of loss declines. Finally, the secular production of heat from the long-lived radioactive substances must come to equilibrate the rate of loss. At this stage there can be no further cooling, save in so far as the radioactive atoms themselves become transformed and cease to yield their energy. The surface richness of the earth in uranium by no means suggests that the earth is now in the process of living down the radioactive heat-supplies by its own original stores of heat. Thus in making the assumption that the earth is to-day radiating into space the heat derived from the atomic systems being transformed in the rocks, we are but claiming for the present a status which, if not now prevailing, inevitably awaits the future.

The uranium which thus appears to make good the heat-loss of the earth is very small—relatively—in amount. The quantity is easily arrived at.

It has already been shown that we can calculate from the gradient and conductivity the average amount of heat escaping from the earth. The result arrived at assigned to this quantity the value 125×10^{-8} calories per square centimetre per second.

Now, one gram of radium involves the liberation of $5{\cdot}6 \times 10^{-2}$ calories per second. Hence, if the whole of the earth-heat came from radium and its radioactive relatives, the maximum quantity of radium present is the quotient of the heat-loss of the earth by the above radiothermal constant. Dividing we get $2{\cdot}2 \times 10^{-5}$ grams. This is the amount of radium which, on the assumptions, must be supposed to exist beneath each square centimetre of the earth's surface inwards. This calculated quantity of radium will, of course, vary with our assumptions as to the thermal gradient. The very steep gradients accepted by some geologists might make the quantity as much as 3×10^{-5} grams per square centimetre inwards.

The parent uranium bears to this quantity of radium the ratio of 1 to $3{\cdot}4 \times 10^{-7}$: the latter figure being the amount of radium in equilibrium with one gram of uranium. Hence, from 70 to 90 grams of uranium beneath each square centimetre of the earth, inwards, is the greatest quantity of uranium we could assume. It will be found later that even this is an excessive estimate, for the *entire* output of heat can in no case be taken as a measure of the uranium and radium within.

When, next, we seek to find the distribution of these radioactive substances inwards, we are met by difficulties. The existence of a specially rich surface layer is easily demonstrated. Its probable downward extension we will discuss later. We shall also find from closer study of the questions involved that there must be uranium and radium deep beneath this layer, probably in very feeble concentration. But this refers only to depths of a couple of hundred miles or thereabouts, and tells us nothing about the state of the great mass within.

The only method now available of approaching the subject of the earth's interior thermal state appears to be that of Lord Kelvin. Unfortunately the knowledge we derive from it is of a negative kind. We learn that we are barred, and, perhaps, barred for ever, from a certain knowledge of the state of the interior. Strange to say, this fact in itself may furnish us with an answer, not certain but probable, to our inquiry concerning the existence of radioactive substances within.

It is not difficult to imagine, even without mathematics, that the movement inwards of the low temperatures prevailing near the surface must be a process of extreme slowness. The distances are enormous, and the feeble conductivity offers a highly retardative mode of thermal escape. We have to imagine columns of material more than comparable with the width of the Atlantic, cooled at one end and throughout charged with great stores of heat. How long must be required for the slowly advancing process of cooling to creep along on its way, nearly four thousand miles extended, the rate of escape diminishing year by year as the surface gradient falls! Immense periods must obviously be involved.

Kelvin showed that if the interior materials of the earth possess approximately the same average thermal conductivity as the surface rocks, and the same average specific heat, even after one thousand million years of free surface-cooling, there would be no sensible loss of heat from an interior core equal to about half the bulk of the earth—the calculation revealing that the entire loss of heat during that long period would be made good from an outer shell 568 miles in thickness; the temperature below that depth not being appreciably less than at the beginning.[1]

It has been argued that the conductivity of the interior materials of the earth may differ greatly from those of which we have knowledge at the surface.[2] And, doubtless, so far as the materials are metallic, there would be greater conductivity within, but so far as they are stony—and it must be remembered that the decisive factors here are the conductivities and specific heats prevailing in the outer 500 miles—there does not appear to be evidence forthcoming in favour of increased conductivity under conditions of higher temperature; indeed, the evidence appears to be the other way (p. 74). The experimental data are, it must be admitted, hardly sufficient to permit any final discussion of the question. When, however, the fact that the probable geological age is but a fraction of the thousand million years is taken into account, it appears improbable that the essential teaching of Kelvin's results will be set aside on the physical constants involved. The extreme slowness, relatively

[1] See also Davison, *Proc. R. S.*, 1887, xlii., p. 325, where the conclusion is reached that after 174,240,000 years the rate of cooling is practically insensible at a depth of 400 miles.

[2] Perry, *Nature*, li., 1895, p. 585.

to the duration of geological time, with which heat would escape from the central parts, seems certainly established. Not only has the mathematical correctness of Kelvin's treatment of the subject never been impugned, but the amendment of his data, whether as regards thermal conductivity, specific heat, density, or assumed initial temperature, seems unlikely to upset his conclusions, stated generally, respecting the relations of the surface with the interior. This general result, which in the original memoir was but a step towards the investigation of the age of the earth, is of much importance to our present views as to the physical state of the earth's interior.

The physical conditions which thus limit our knowledge of the earth's interior may, as stated above, in themselves involve a solution of the problem of the inner thermal state.

Obviously if uranium exists in the thermally isolated interior, these regions must now be rising in temperature. A little thought will show that Kelvin's result really removes the only argument that might be urged against such a condition of things. It may have been so for many millions of years—since in fact, convective conditions came to an end—and yet have produced no exterior effect, nor interfered with the secular cooling of the earth's surface which mountain-building phenomena indicate as having progressed throughout geological time. Let us assume for a moment that radium was contained in the interior materials of the earth in quantities comparable, say, with those prevailing in ocean waters: say $0{\cdot}02 \times 10^{-12}$ grams per gram. Then it is easy to show, by multiplying this value by the heat out-put of radium per second and by the time in seconds, that in a hundred million years only 3½ calories are evolved per gram; so that if the specific heat be taken as one-fifth that of water, the rise of temperature is only 17·9 degrees. Evidently a much more considerable amount of radium might be present, and the resulting temperature-rise during 10^8 years be such as would probably produce no surface effects.

This of course leads us to regard the question of the existence of interior uranium as an important one: and we turn with some interest to the arguments for or against the internal radioactivity of the earth. There are many and weighty reasons (so far as arguments on probability and analogy can be regarded as weighty) for believing that there is interior uranium in the earth.

(*a*) Its presence at the surface suggests that it is an original terrestrial constituent. It is in the highest degree improbable that an element can be entirely restricted to one part of the earth, and not exist in some degree in every part. Ultimately, indeed, it is probably true that every element is everywhere, and would be so perceived had we but the sufficiently delicate means of appreciation. Apart from this concession to the number and minuteness of atoms, it is unlikely that so considerable a quantity of the element uranium should occur in the surface rocks, no matter how deep-seated their origin, and not exist far into the interior, if not to the very centre. This argument based on the principle of uniformity is only met by assuming that uranium is not a terrestrial constituent at all; that it was not primevally upon the earth, but was picked up from space during the long ages of geological time. This view may involve no impossibility, but, in the first place, the internal materials, if ever they circulated in meteoric form around the parent star, should also have gathered the radioactive element, whether cosmic in origin or emanating from the sun. And in this way it should be incorporated in the internal materials of the world as finally evolved. Secondly, this theory is unsupported by any exceptional radioactivity either in meteorites or in cosmic particles gathered from the ocean floor.

(*b*) A physical explanation can be offered of the surface concentration of uranium (to be referred to later), but only on the assumption that it came from the interior, or at least from deep beneath the crust. Thus the presence of uranium at the surface—if the suggested explanation be correct—is an indication of the interior presence of the element, as the surface ore tells the miner of the metal beneath.

(*c*) Meteorites are regarded with some probability as representing terrestrial materials not accessible to us at the surface. Such observations as exist on their radioactivity have shown small quantities of radium in the stony meteorites and highly nickeliferous æosiderites, but none in meteoric iron. However, it is present in metallic iron derived from the earth's interior.[1]

(*d*) So long as we regard the earth as a derivative

[1] Strutt, *Proc. R. S.* 77 A, p. 480.

of the solar mass, we cannot logically deny a similarity of constituents. Now, there is good evidence for the presence of uranium in the sun. There is evidence from the spectroscope, and evidence from the abundance of helium in the solar atmosphere: helium being upon the earth unquestionably a product of radioactive change, in part, if not entirely. There is also the evidence based on the difficulty of accounting for the long continuance of solar heat without the help of radioactive contributions; indeed, the evidence from helium and from the necessities of solar theory would lead us to suppose that uranium was a very considerable constituent of the sun. True, we cannot suppose such a quantity to be present as would account for solar heat in its entirety, for it has been shown that the amount of heat emitted by the sun every second would require as much radium in each unit of solar mass as exists in pitchblende. Now, the radium-richness of pitchblende is dependent on the 60 per cent. (or thereabouts) of uranium present. It seems impossible to suppose that so great a proportion of uranium can exist in the sun. Still it remains a fact that the evidence is in favour of regarding uranium as a solar constituent. And as such, consistently with theories of the earth's genesis, we regard it as an original terrestrial constituent.

This array of argument for the presence of uranium within the earth might still be deemed inconclusive if there was any strong case for the opposite view; but, in a word, there is no case for the other side. For we really possess no criterion in our surface observations as to whether or not the interior is rising in temperature. We know nothing of how the effects of even a considerable rise might be kept under control. Every event of geological history might have gone on as we believe it to have occurred, and no sign be manifest of accumulating energy within the central core.

Thus the conclusion that the earth's interior temperature is rising seems the logical outcome of the probable conditions taken along with Kelvin's proof that the central parts of the earth are thermally isolated for immeasurable periods of time from the surface. Unless we take refuge in arbitrary denial of the existence of the radioactive elements within, or assume that in some manner the radioactive transformation of uranium is annulled by the conditions of pressure or temperature, we seem to have no option but to accept this conclusion as the answer to our inquiry into the state of the earth's interior.

The question as to what must ultimately result from a radioactive production of heat and its storage in the interior of the earth seems beyond our power to determine. All must, of course, depend upon the amount of uranium and radioactive elements contributing to the accumulating heat. If the supply is rapid, the effects must ultimately be felt at the surface; if slow enough, the intervention of another factor has to be considered: the decay and gradual disappearance of the radioactive materials themselves. These, under certain conditions of radiothermal supply, must entirely perish, and the accumulation of heat come to an end, before any surface disturbance is produced. The collected heat would, in this case, slowly escape in remote future times, unattended by any effects upon the crust. This will be understood when the maximum amount of accumulated heat is estimated for a particular case. Thus, if we assume, as before, a quantity of radium $0{\cdot}02 \times 10^{-12}$ grams per gram for the central parts of the earth, and the quantity of uranium present per gram be taken as roughly a measure of the total potential radium, we have:—

$$\frac{0{\cdot}02 \times 10^{-12}}{3{\cdot}4 \times 10^{-7}} = 6 \times 10^{-8}$$

grams of radium (uranium) per gram. (The denominator of the fraction will be recognized as the quantity of radium in equilibrium with a gram of uranium.) The total quantity of heat which is involved in the transformation of one gram of uranium, reckoned on the average life (2540 years) of the potential radium, is $4{\cdot}5 \times 10^{9}$ calories. Hence the total heat during the whole life of the uranium derived from materials of the radium-richness assumed is $6 \times 10^{-8} \times 4{\cdot}5 \times 10^{9} = 270$ calories per gram; and if the specific heat is one-fifth that of water, the temperature attained is 1350°.

This result assumes that during the whole time of decay of the uranium there is no escape of heat, which would hardly be the case: the greatest rise of temperature would probably be much less. In short, it is possible that a uranium-richness of this order would not result in a disturbance of the crust.

But if we assumed that there was 100 times as much uranium—that is, that the radioactivity was 2—then it is easy to show that the stored heat accumulating in 100 million years would give rise to the temperature of 1800°; that is, it would increase the initial temperature by this amount. And on Kelvin's investigation we see that this great quantity of energy might to a large extent be retained. Where such temperatures are involved it is probable that before the hundred million years had run their course the world, with such a constitution, must become once again a heated mass throughout.

We possess really no assurance that such a consummation does not await the future, nor evidence that such has not more than once been an event of the past.

A celestial body possessing any considerable store of long-lived radioactive elements will not cool as an ordinary body, simply parting with its stored-up heat, and falling in temperature gradually from its surface inwards. Suppose it starts on its career at a temperature of liquefaction. The rate of cooling will then be great, and will only sensibly be delayed by the dynamical heat generated by its shrinkage upon itself. But this supply, too, will gradually become exhausted, a crust will form, and if Kelvin's ideas are right, a *consistentior status* will inaugurate a geological age. But the rate of cooling has now become relatively insignificant; probably not the ten-thousandth part of what prevailed before the *consistentior status.* The quiet accumulation of radioactive energy proceeding throughout the mass will, near the surface, make good the radiation loss, but in the interior, where no means of escape exists, must collect during the passing geological periods.

There can only be one result:—general surface vulcanicity and reversion to temperature conditions which may involve the repetition of the entire sequence of events.

If such has been the past history of our globe and such the origin of our geological age, we find the existing earth-heat to be more or less completely radioactive in origin. But this invalidates nothing of what we have previously stated as to radium-heating reconciling estimates of the earth's geological age. The earth-heat might be entirely radioactive in origin, and yet the quantities of radioactive materials involved might be no more than would preserve the present surface conditions for many millions of years.

The beforementioned possibilities must be clearly distinguished from the view that the existing state of equilibrium between radioactive supply and loss by radiation also obtained at a period when the earth was radiating many thousands of times its present amount of heat. For in this latter case we necessarily assume the former existence of thousands of times the present amount of radioactive materials; and, if we mean uranium, we have to put back the *consistentior status* such a time-interval as will suffice for the decay of this uranium to its present small quantity. To multiply the existing uranium 4000 times requires us to revert to a past 60,000 million years ago. Such a lapse of time since the consolidation of the earth or of its surface brings us into hopeless difficulties not only with reference to geological time, but with regard to solar theory.

We have no reason to find in the advent of radioactivity into cosmic and geological sciences any new difficulties, but rather the clearing away of old ones. The duration of solar heat finds a plausible explanation, and the prolonged duration of geological time without the corresponding cooling of the earth's surface becomes accounted for. But it cannot be said that it enables us to foretell the destiny of our globe. Unless other lines of evidence exist, we are left to the imagination as to what has been the history of the past, and what may result in the future. But wonderful possibilities are brought before us. Peaceful cooling may await the earth or catastrophic heating may lead in a new era of life. Our geological age may have been preceded by other ages, every trace of which has perished in the regeneration which has heralded our own.

Whatever be the future or the past of our world, we have the untrammelled regions of space in which such varied destinies must surely find their accomplishment. The planets may now be in varying phases of such great events. And when a star appears in the heavens where before we knew of none, may not this be a manifestation of the power of the infinitely little over the infinitely great—the unending flow of energy from the unstable atoms wrecking the stability of a world?

CHAPTER IX.

RADIOACTIVITY IN THE EXTERIOR PARTS OF THE EARTH.

INDICATIONS OF HIGH SUB-CRUSTAL TEMPERATURES—THICKNESS OF THE CRUST—LIMITING DEPTH OF THE RADIOACTIVE LAYER—ALLOWANCE FOR COOLING OF THE EARTH BY DECAY OF THE URANIUM—NECESSITY OF A DEEPER AND LESS CONCENTRATED DISTRIBUTION OF THE URANIUM—APPROXIMATION TO THE THICKNESS OF THE RADIOACTIVE LAYER—SUGGESTED ORIGIN OF THE SURFACE CONCENTRATION OF THE URANIUM.

AMONG the surface materials of the earth we everywhere find manifestations of deep-seated high temperature. It may, indeed, be said that almost every part of the earth has been at some time visited by volcanic outbreaks. And where evidences of such are no longer obtainable, we are sure that it only requires the disclosing effects of denudation to show the buried igneous masses—some solidified deep beneath the outer rocks, others brought higher up on their way to the surface. Again, outflows, not strictly volcanic, have flooded great areas to the extent of thousands of square miles. No more convincing evidence of internal heat could be required, and indeed it is hard to imagine any additional evidence of the existence of fusion temperatures beneath.

The reasons for supposing that radioactivity has been and is instrumental in sustaining the temperature of the earth have been given in the last chapter. Following on this, we see that, whatever conclusions we now adopt respecting the distribution of uranium near the surface of the earth, we necessarily conclude it must be such as will have sustained the intense temperatures beneath the crust during a large part of geological time.

Differences of opinion, comparatively unimportant, exist as to what precise temperatures may prevail beneath the crust. The only guide we appear to have in the matter is the temperature of melted lava flowing out at the surface. This has been estimated at from 1000° to 1300°, and even to 1400°. There are reasons for believing that 1300° is the least temperature we can ascribe to basaltic magmas deep within, or just below, the crust.[1]

The thickness of the crust is an indeterminate and, in any case, a somewhat indefinite quantity. If we mean of the rigid crust, we first require to know the temperature at which rigidity fails under the prevailing conditions of stress. There appears to be no abrupt beginning of plasticity in a siliceous material when its temperature is gradually raised. The range of plasticity is considerable (especially in the more acid rocks), and long-continued and intense stress will affect siliceous materials at temperatures considerably below the melting-point. We are, however, not concerned here with the precise distribution of temperature and rigidity. It is sufficient to state that a full fusion temperature is, by some writers, believed to prevail at about 38 kilometres from the surface; others would refer such temperatures to greater depths. What has

[1] Daly, *Am. Journ. of Science*, xxvi., 1908, p. 35.

already been assumed, in relation to the subject of the stability of the crust, would place the true fusion-point some few kilometres lower down. It is probable that the thickness of the crust is variable, within limits, over the earth, and, as we have seen, may be considerably reduced beneath the areas of deep sedimentation.

These views define, to some extent, the limitations which must guide us in dealing with the question of surface radioactivity.

The most immediately important conclusion arising from the investigation of the radioactivity of the surface-rocks is, that a layer of richly radioactive materials must cover the earth, which can only extend but a few miles downwards. This is easily demonstrated. The proof is based on the known amount of heat escaping from the crust and being continually radiated into space. We have seen in the last chapter that if we refer, without deductions of any kind, the whole of the escaping heat to a radiothermal origin, we can have no more than from $2{\cdot}2 \times 10^{-5}$ to 3×10^{-5} grams of radium beneath each square centimetre of the earth's surface. Now, if we take it that 3×10^{-12} grams per gram, or say, 8×10^{-12} grams per cubic centimetre, represent the average radioactivity of the igneous rocks, it is evident we can estimate the depth to which rocks of this richness can extend by a simple division of the maximum quantity of radium by the amount in each cubic centimetre. This gives us a limiting depth of from 27·5 to 37·5 kilometres (17·2 to 23·5 miles) according to the gradient we prefer.

Hence if we take the surface-richness as indicative of what is beneath, the radium must entirely cease at these depths.

This simple calculation, however, fails to represent the actual facts. We find on examination that neither can we suppose the thickness of the radium-rich layer to be so great as that which we have arrived at, nor can we even suppose that all the radium is collected in the assumed manner at the surface. Hence we have to modify the estimate of this layer as arrived at above. The necessity of doing this must now be shown.

In the first place, we based our calculation on the quantity of heat escaping at the surface, assuming that this was a measure of the radium. But is it? The parent substance, uranium, is continually waning in amount. True, the process is very slow, but still it is ever diminishing according as it becomes transformed: at a rate of about the ten thousand-millionth of its amount in a year. But by assumption this substance is responsible for the maintenance of the temperature of the earth. It may even, as we have seen, lead to a rising temperature within. In the exterior parts it acts to make good the surface loss. Hence, as it diminishes in quantity, the temperature must fall. Thus the fall in temperature of those parts of the earth which send heat to the surface, and are kept hot by the uranium, will just be the ten thousand-millionth part of their temperature each year. This is admitting, of course, that the earth is not, at least at the surface, preserved at a strictly constant temperature, but is cooling. A fall in temperature, however, involves the escape of some of the existing store of heat. That is, if the materials fall t degrees, there must pass out a quantity of heat which is found by multiplying t by the specific heat, and by the mass which is cooled. Plainly the importance of the heat-loss will, in this case, depend mainly on the mass which is cooling. Suppose the whole earth was cooling, the conductivity being sufficiently good to permit an almost uniform drain of heat from every part of the interior; then the number of calories annually escaping at the surface becomes very great, in spite of the minuteness of the drop in temperature. Thus, if we take the mean temperature of the earth as, say, 1500°, the annual fall will be 10^{-10} of this, using here a rather low value for the transformation constant of uranium. We have also the mass as 6×10^{27} grams. The average specific heat may approximate to one-fifth that of water. The quantity of escaping heat due to cooling is then

$$1500° \times 10^{-10} \times 6 \times 10^{27} \times 0{\cdot}2 = 1{\cdot}8 \times 10^{20} \text{ calories.}$$

But we cannot assume the total heat escaping from the earth, according to the gradients, as more than $2{\cdot}6 \times 10^{20}$ calories per annum. So that we are left with but $0{\cdot}8 \times 10^{20}$ calories per annum as a measure of the radium actually contributing heat to the surface: i.e., less than a third part of the quantity we assume if we neglect the fact of the decay of the uranium. We can (partially) escape this conclusion in only one way; by recalling the fact that the whole mass of the earth does not send heat to the surface, but a large part of it must be accumulating what radioactive heat is generated therein. For this interior part the decay of the uranium only involves a lessened rate of accumulation. From this we see that 6×10^{27} is too large a value to allow as representing the cooling mass sending heat to the surface; and hence the deduction for earth-cooling must be less.

The principles assumed in the foregoing argument are perhaps not easily followed by those unaccustomed to calculations involving thermal measurements. An analogous case may help. Let us take the familiar instance of a boiler heated by a furnace within. The heat of combustion we may suppose diminishes a little owing to gradual consumption of the fuel, and the temperature of the whole mass, boiler and its contents falls by a corresponding amount. Now, it cannot do this without parting with some of its stored-up heat. This must escape by radiation at the surface; it can go nowhere else. The amount of this heat in calories is calculated on the average specific heat (the average quantity of heat which is liberated when a gram of the materials falls one degree in temperature), multiplied by the fall in temperature. The quantity of heat thus arrived at as passing out at the surface might, by an observer without, be confounded with the continued heat-production of the furnace. In fact, the observer of the escaping heat is liable to overestimate the activity of the combustion unless he can also observe the fall in temperature. But this might be too small to be detected, and yet if the mass of the boiler and its contents is great enough, the error might be a very large one. So it is in the case of the earth; the temperature-fall is quite beyond detection, but, owing to the enormous mass, the heat of cooling as distinct from the radioactive heating may be very large.

The difficulty in determining the true correction for cooling in the case of the earth is similar in nature to that which we met before when discussing Kelvin's results as to the interior temperature, and obviously arises from the same source—our ignorance of the physical properties of the materials far down. If at a venture we assume that one quarter of the earth's mass is falling in temperature with the decay of the uranium, then we have the following figures. On the 32-metre gradient the total annual loss of heat from the whole earth is 2×10^{20} calories, and the heat referable to falling temperature is $0{\cdot}45 \times 10^{20}$ calories: hence there remains $1{\cdot}55 \times 10^{20}$ calories as the heat derived from continued radioactivity.

This necessarily entails a lesser thickness for the radioactive layer than that with which we started. We shall next see that even the quantity of radium to which we would be limited by this calculation is in excess of what we can assign to the surface layer. This arises from the requirements as regards temperature of the deeper parts of the crust.

On the evidence there must be temperatures of 1200° to 1400° prevailing beneath the crust. Now, if the sub-crustal temperature is indeed maintained by the uranium, then the radiothermal actions must be adequate to give rise to these temperatures, or we must look to some other source of heat—say, the original heat of the earth—to explain the high temperatures. The arguments against having recourse to the original heat of the earth have already been given. We therefore agree to the necessity of so distributing the radioactive materials in the exterior parts of the earth that the said temperatures may be attained. As a preliminary consideration it is plain that if the uranium be all placed near the surface, lesser temperatures will be attained by radiothermal actions than if it, or a part of it, be more deeply buried; for in the first case the easy escape of the heat would forbid the attainment of a high temperature. And this is just what would happen if all the radium in the outer layer were at a mean concentration of 8×10^{-12} grams per cubic centimetre of rock. Such a layer would extend downwards a distance defined by the assumption already arrived at, that $1{\cdot}55 \times 10^{20}$ calories per annum may be assigned to the radiothermal actions. This, however, involves the uncertain correction for the heat of cooling of a portion of the earth. To more effectually establish the present point, let us neglect this correction entirely, and assume the radioactive materials to extend to 27·5 or 37·5 kilometres, as first calculated.

We are, it must be remembered, going on the hypothesis that the earth derives its escaping heat *and its temperature* from this radium jacket. The loss of heat has, in fact, guided us in calculating the thickness of the jacket, and now we seek to find if this layer can give rise to a sufficient sub-crustal temperature. The calculation of the maximum temperature to which it can give rise is not difficult. No heat enters it from below, for the temperature beneath is supposed to depend on the basal temperature of the layer, and to be in equilibrium with it. Heat, therefore, only flows out above as fast as it is generated in the 27·5 or 37·5 kilometres of radioactive materials.

The equation given in a previous chapter (p. 77) enables the basal temperature to be found[1]:—

$$\theta = \frac{qhD^2}{2K},$$

where q is the radium content of a cubic centimetre of the layer, D its thickness in centimetres, h the heat generated per second by one gram of radium, and K the thermal conductivity throughout the layer. We have q and D as above. The value of h already given more than once is $5{\cdot}6 \times 10^{-2}$, and K we take as 4×10^{-3}. Putting in these figures, we find that the basal temperature is but 420° if D is $27{\cdot}5 \times 10^5$, and it is 790° if D is $37{\cdot}5 \times 10^5$.

The reason for this failure of the radium to give the necessary high temperature is simply that we have placed it too near the surface, so that its heat too readily escapes. By supposing a sufficiently sparse distribution of it in the upper layer we may give the layer a greater depth, and so raise the basal temperature; but then we appear to abandon the indications of the surface rocks, and admit that they do not indicate the radioactivity throughout the jacket.

And this is what we really are driven to do. The

[1] See Appendix A.

calculation shows that the immediate surface distribution cannot indicate a condition continued downwards till all the allowable radium is exhausted. There must be radium of very great dilution extending downwards beneath the immediate surface materials to such depths as will maintain the high temperatures beneath the crust. Only by assigning to the lower parts of the crust and the sub-crustal magmas a feeble radium-content, considerably less than we perceive at the surface, can we bring our views into harmony with the temperature conditions which we know to prevail beneath the crust.

Stated in terms so general this conclusion is probably a safe one; but when we try to make it more definite, uncertainties thicken around us. The requirements of our surface observations may, however, guide us to some idea of the depth to which the radium-richness observed in the igneous rocks probably extends.

The intruded and extruded igneous rocks appear to show no diminution in radioactivity. They must probably represent general conditions prevailing far down. Not necessarily so deep as the zones of general fusion, however, for, as we have seen (Chapter V), where the phenomenon of igneous intrusion is most evident, the sub-crustal geotherms have been raised from their normal positions, so that re-fusion and outpouring of materials, which elsewhere might be located above the fusion zone, are brought about. Still we here find difficulties if we assume the higher radioactivity to prevail but a little way down; and, on the other hand, we must abandon the view that the earth's temperature is maintained by radium if we concentrate radioactive substances entirely in the upper parts of the crust.

Now, taking the gradient of 32 metres as the best approximation we can make to a mean gradient, we have 27·5 kilometres of crust having the radioactivity of 3 as a manifestly excessive estimate of the thickness. We have to reduce it in consequence of some of the earth's escaping heat being derived from cooling by decay of the uranium, and again we have to reduce it because a considerable amount of uranium must be located further from the surface. We do not know what to allow for either of these retrenchments. If we said that the richer part of the crust must be between 9 and 15 kilometres deep, we cannot be far from the truth. This appears to be the best we can do on our present knowledge. Twelve kilometres have been taken for purposes of calculation in Chapter V.

The surface concentration of a part of the uranium is a fact beyond question. At first sight it appears as if this was just what we would not have expected. Uranium is the heaviest element known, and a central concentration would rather have met our expectations. However, a physical cause can be assigned which does not appear to conflict with those views of earth-genesis which involve the formation of the globe by meteoric accretion.

The suggested explanation follows easily if we admit the possibility that the uranium contained in the outer shell of the earth—say, to a depth of a few hundred miles—was not originally uniformly distributed, but richer here and there in a manner that might naturally arise on any meteoric or "planetesimal" theory of the earth's origin.[1] Thus the constituents might have been derived from different depths of the solar mass, so that when finally aggregated to form our earth, they would give rise to local variations of material. We might say simply, without theorising, that they possessed the chemical differences observed in meteoric materials. The consequences arising from this state of things are readily seen.

There would be a certain temperature ensuing from dynamical effects which would originally prevail throughout these aggregates, and which would be independent of their uranium content. But, as time went on, the materials richest in uranium would acquire the highest temperatures; and when slow convective motions were possible, such hotter masses would tend to move away from the centre of the earth under the force of displacement arising out of their lessened density—just as a region of heated water beneath the ocean must slowly rise to the surface. Such convective actions may have taken place principally in the early stages of terrestrial heating, and possibly may have been restricted to limited radial displacements; but, evidently, if they involved the transport of materials even for a very few hundred miles, and took place near the surface, the result must be the

[1] Chamberlin and Salisbury, *Geology*, vol. ii.

accumulation at the surface of the more uranium-rich materials contained in the outer parts of the heated earth. And once these attain the surface, there are afterwards no forces to cause them to retrace their path.

Again, there is no improbability that such movements, acting with extreme slowness over geological time, and requiring long periods for the accumulation of radioactive heat, might affect the already formed crust of later geological times. Some of the great lava flows of comparatively recent date may be associated with such actions.

It considerably strengthens this view of the cause of the radioactive surface layer if we consider as possible cases some of the great lava flows which have broken through the crust in comparatively recent times. The most striking instance is that of the Deccan basalts and traps, which, towards the close of Cretaceous times, were poured out, covering an area of 200,000 square miles to an average depth of about one mile. This volume of basalt would fill a sphere 115 kilometres (72 miles) in diameter. According to the observations quoted in Chapter III, the average radioactivity appears to be about 14×10^{-12} per cubic centimetre. Suppose such a mass, situated 100 or more miles beneath the surface, originally at the temperature and density of its surroundings, and now left to heat up under the radioactive energy. If we assume that over the surface—approximately spherical—heat freely escapes, so that the surface does not rise in temperature above the surrounding magma, then the equation giving the final excess of temperature at any distance r from the centre is[1]:—

$$\theta = \frac{qh}{6K}(R^2 - r^2),$$

where q is the radioactivity, h the heat constant for radium, and K the conductivity. These constants have previously been given.

This gives at the centre an excess temperature above the surroundings of 1070°. The mean temperature would, of course, be less. At a distance inwards from the surface of the sphere of 28 kilometres—that is, 30 kilometres from the centre—the excess temperature is 784°. These temperatures would be lower if the shape was not spherical, but would certainly be higher under the actual surface conditions; for the escape of heat would not be perfect, but the surrounding magma would tend to shut in the heat, being itself a bad conductor. Indeed, the viscous state of this surrounding magma might be conditioned by the heat generated in the radioactive materials. If we suppose that some mean excess of temperature of about 500° can be ascribed to the whole mass (and convective effects would interfere with the attainment of a very high temperature at the centre), then we can calculate approximately what the force of buoyancy, which acts to urge the whole to the surface, would be.

Barus gives the coefficient of expansion of liquid diabase from 1100° to 1500° as 0·0000470; so that heated through 500°, one cubic centimetre becomes

[1] See Appendix A.

1·0235 cubic centimetres; or the gain in volume per cubic centimetre is 0·0235 cubic centimetres. But the density of diabase at its melting-point (1094°) is from the data given by Barus deducible as about 2·62. If this be taken as the density of the surrounding magma, then the increase in volume of the superheated mass involves a displacement force of 0·061 gram, or about 60 dynes, per cubic centimetre. We may put this more briefly by saying that the volume increases about 2·3 per cent., due to the rise of temperature, and that, therefore, an upward-directed force, amounting to 2·3 per cent. of the weight of the displaced magma, acts to bring the radioactive mass to the surface. It is probable that effects of this kind would be restricted to conditions of high viscosity, and that convective movements would be extremely slow. The calculation may, however, give some idea of the magnitude of the forces involved. The general outcome of such displacements of the more radioactive nuclei to the surface must be the impoverishment of the depths and the corresponding enrichment of the upper crust.

CHAPTER X.

THE RADIOACTIVITY OF THE ATMOSPHERE.

EFFECTS OF EMANATION IN THE ATMOSPHERE—IONIZATION OF GASES NOT SPONTANEOUS—NUMBER OF IONS IN THE ATMOSPHERE—RATE OF PRODUCTION AND RECOMBINATION—GEOLOGICAL EFFECTS OF THE IONS—ACTIVE DEPOSIT OBTAINED FROM THE AIR—GENERAL PREVALENCE OF THE ACTIVE DEPOSIT AND DEPENDENCE ON METEOROLOGICAL CONDITIONS—ACTIVE DEPOSIT FROM RADIUM EMANATION—AMOUNT OF RADIUM EMANATION IN ONE CUBIC METRE OF AIR—EXPERIMENTAL RESULTS ON EFFECTS OF EMANATION IN THE ATMOSPHERE—CALCULATED EFFECTS OF THE EMANATION—PERCENTAGE LEAK OF EMANATION FROM THE SOILS—IONIZATION OVER THE OCEAN—EFFECTS OF PENETRATING RADIATIONS FROM THE EARTH — ESTIMATE OF THE RADIOACTIVITY OF THE SURFACE ROCKS—PRESENCE OF THORIUM EMANATION IN THE ATMOSPHERE—EFFECTS OF RADIOACTIVITY ON THE CHEMICAL COMPOSITION OF THE AIR.

THE facts recorded in Chapter III show that radioactive substances are to be found in small quantities not only in the rocks but in the ocean and rivers. We have also seen that the transformations of parent radioactive elements give rise in every case to gaseous elements—the emanations of radium, thorium, and actinium—which, after a short period of existence, lose a part of their mass and re-transform to solid substances. The solids which result from the transformation of the emanations are among the most unstable of the radioactive elements; they possess periods of existence measured by minutes or hours, and are then succeeded by generations as unstable as themselves; so that a considerable part of the total atomic evolution is comprised within some minutes or hours succeeding the break-up of the emanation. The fact that a gaseous element arises in the radioactive sequence suggests the probability of radioactive energy not being confined to the solid and liquid substances of the earth, but extending to the gases collected around the globe. If the gaseous radioactive elements escaped into the atmosphere, it is to be expected that the succession of changes which succeed the gaseous state must exert effects similar to those arising in the electroscope when emanation is admitted: that is, they must ionize the air, rendering it thereby a conductor of electricity.

We have already referred to the theory by which it is sought to explain the electric conductivity of gases; and, incidentally, to the fact that the air in the electroscope possesses a certain amount of conductivity even when no radioactive gas has been intentionally introduced. This persistent conductivity of air—and indeed of all gases—was by the earlier observers referred to causes which are now known to be inadequate. Within recent years, and mainly in consequence of the work of Sir J. J. Thomson and his pupils, the conductivity of gases has found a more consistent explanation in the carriage of elemental electric charges of both signs, by certain ionized atoms or particles in the gas; the gas itself being, when free from these carriers, a non-conductor.

This view must not be understood as in all respects similar to the much earlier theory of the electrolytic conductivity of liquids containing salts in solution. We might refer to important points of distinction between the two hypotheses: one of these claims our special attention. The ionization within the salt solution arises from internal conditions and the interaction of internal forces. It is an inherent property of the solution, and the number of ions present in unit volume is controlled mainly by the concentration of the dissolved substance. Even when no salt is present, and we have to do with the solvent only, a certain amount of ionization exists arising from internal conditions which we do not seem able to control. It appears to be otherwise with gases. There seems to be no reason to believe that any inherent specific ionization exists, although internal factors influence the electric state. Certain it is that we can largely control the electric conductivity within a closed vessel by taking precautions, as to (*a*) the freedom of the enclosed air from any trace of radioactive emanation; (*b*) the use in the construction of the vessel of materials which are free from radioactive elements; and (*c*) the protection of the enclosed air from penetrating radiation from without. Now, it may be said that not one of these precautions can be applied with any assurance of completeness. The retention in every case of a certain amount of conductivity need not, therefore, cause us to modify the view that the ionization of a gas is something impressed upon it from without.

In addition to a host of experiments showing how largely the applied external conditions control the ionization of a gas in a closed vessel, the striking fact has been observed that the ultimate residual ionization is proportional to the density of the particular gas. This observation has been found to apply even to gases of complex molecular structure and much higher in density than air.[1] This seems to show that the ionization observed must be due to a radiation either from the walls of the vessel or from external sources. For it would be expected that effects arising from radiated energy would be more manifest in gases wherein a high density favours the more complete absorption of the energy.

The atmosphere possesses normally a variable number of ions which may be from 1000 to 3000 of each electric sign per cubic centimetre, and may even vary considerably outside those limits. Thus Eve obtained at Montreal, in hazy, hot weather, so few as 370 positive and 367 negative ions per cubic centimetre. The numbers seem to be affected by atmospheric conditions to a certain extent. Dust or haze in the air appears to facilitate the combination of the ions, and so to reduce their numbers. There are generally found to be a larger number of positive than of negative ions, the ratio being sometimes so great as 1·5 to 1.

As the recombination of the ions is continually going on, it is evident that there must be a perpetual supply to make good the losses. This at once follows from the principles of the ionic theory. Now, it can be shown that the relation between the number q, of ions of one sign, brought into existence in unit volume and in unit time, with the number, n, present when

[1] Rutherford, *Radioactivity*, p. 535.

there is equilibrium between production and recombination, may be represented by the equation $q = \alpha n^2$, where α is the coefficient of recombination; a relation expressing the fact that the number recombining is proportional to the square of the number present.

The determination of the number present at any instant may be effected by observing the rate of loss of charge of an insulated electrified body immersed in a current of the gas; the conditions being those of saturation: that is to say, the potential being such that all the ions present are utilized to discharge the electric system before appreciable recombination can occur. These conditions do not seem to be, in practice, perfectly accomplished. Certain slow-moving ions have been shown to be present in the atmosphere which will not be detected by this method. In these determinations the charge upon the ion necessarily enters into the calculation; and as Rutherford and Geiger[1] have recently shown that the value previously ascribed to the charge—$3{\cdot}4 \times 10^{-10}$ E.S. unit—should be raised to $4{\cdot}65 \times 10^{-10}$, the estimates generally given of the number of ions in the atmosphere must be reduced in the ratio $4{\cdot}65 : 3{\cdot}4$.

The value of α for dust-free air has been determined by Townsend[2] as equal to $3420 \times e$, where e is the ionic charge. Taking Rutherford and Geiger's value of e (as above), we have $\alpha = 1{\cdot}6 \times 10^{-6}$. If now we assume 2000 ions present per cubic centimetre, and apply the value of α, we get for q 6·4 ions per cubic centimetre per

[1] *Proc. R. S.*, 81 A, p. 170.

[2] Townsend, *Phil. Trans. R. S.*, 81 A, p. 170.

second. However, under ordinary conditions it would appear that q may have a considerably greater value: the rate of recombination being higher in dusty or hazy air. Thus at Manchester, Schuster found that a was two or three times as great as for dust-free air, and that q varied from 12 to 38·5.[1] On the other hand, in the pure air over the ocean, with about 1000 ions per cubic centimetre, the rate of creation of ions would appear to be no more than one or two per cubic centimetre per second. In hazy weather n may fall to very low values, owing to an increased rate of recombination and the absence of any exceptional compensatory ionizing actions. The effects of dust and haze have been explained by Sir J. J. Thomson as arising from the greater area offered by suspended particles when they become electrified by union with an ion.

From these facts it will be seen that before we can confidently assert the adequacy of any source of ionization which we may discover as accounting for the rate of production of ions in the atmosphere, the local, simultaneous conditions must be completely determined. For a may vary not only from place to place, but from day to day in the same locality. And, as will presently be seen, the probable causes which create the ions are themselves variable in character.

The continued creation of large numbers of ions in the atmosphere is the fundamental fact with which we are concerned, and to which we must here confine our attention. Its bearing on geological processes is latent in many observations made within recent years.

[1] Schuster, *Proc. Manchester Phil. Soc.*, p. 488, 1904.

Thus it is found that the ions possess the remarkable property of precipitating water from the vaporous into the liquid state. At the time when Aitken and others first observed the fact that water-vapour mingled in dust-free air refused to be precipitated under conditions of supersaturation, the explanation was thought to reside simply in the absence of dust-particles which might act as nuclei for the initiation of condensation. It is now known that something else is involved. The ions present are also effective in causing condensation. When all dust has been removed, the presence of any source of ionization, such as X rays, or the rays from a radioactive substance, is effective in bringing about precipitation when a state of supersaturation exists. Again, carefully made experiments reveal the fact, that in the absence of any such applied sources of ionization, precipitation will always ensue in dust-free air if the degree of supersaturation exceeds a certain limit: a result pointing to some persistent source of ionization.[1] There can hardly be a doubt that we are justified in applying these facts to the explanation of certain meteorological phenomena.

Throughout a moisture-laden atmosphere these remarkable precipitating actions must progress, in greater or less degree; and where the water-vapour is sufficiently abundant, probably determine its descent in the form of rain, hail, or snow. The meteorological importance of the large numbers of ions normally present in the atmosphere must be great. The physical principles underlying these actions cannot here be enlarged upon.

[1] C. T. R. Wilson, *Phil. Trans.*, A, 189, p. 265.

It seems certain, however, that the electric charge is directly involved in establishing the nuclei required for the initiation of aqueous precipitation.

The last remark leads us to consider another meteorological effect seemingly connected with the ionic precipitation of water from the atmosphere. From experiments made by C. T. R. Wilson[1] it has been established that the precipitation of water-vapour on the negative ions occurs much more readily than upon the positive ions. Under conditions of increasing quantities of aqueous vapour in the atmosphere, precipitation will, accordingly, be first induced by the negative ions, and the rain-drops so initiated will hurry the negative charge to the earth, leaving the positive ionic charge of the atmosphere comparatively little altered in amount. Thus, consider the case of a quantity of air—it might be many cubic miles in volume—which has become charged with water-vapour near the ground, and then is borne to the upper regions of the atmosphere, by its own buoyancy, by the action of air-currents, or by displacement upwards when carried across a mountain range. There will ensue a rapid expansion and cooling; and the conditions obtaining in Wilson's condensation vessels will be reproduced. The first ionic precipitation will be brought about by the ions of negative sign, and their abstraction from the higher atmosphere may ensue. These effects might have been anticipated on the grounds of Wilson's results; that they occur is supported by the fact that the atmosphere is found to possess a prevailing

[1] *Phil. Trans.*, A, 193, p. 289.

positive electrification respecting the surface of the earth. It is possible that the intense electrical phenomena which attend, and follow upon, rapid condensation of water in the upper atmosphere, arise from these subtle causes.

The progress of radioactive science early suggested the inquiry as to whether the ionization in the atmosphere might not be traceable to radioactive energy. The pioneer observations in this direction were made by Elster and Geitel. These observers made the remarkable discovery that definite radioactive substances could be gathered everywhere from the atmosphere, by freely exposing a negatively charged surface. They used a wire raised to a high potential. Upon this the active substances collected, and from this they could again be removed by friction, or by solution in an acid. These substances, when brought into an electroscope, much increased the ionization. The idea was a natural one that the normal ionization of the atmosphere must be, in part at least, ascribable to the presence of these radioactive bodies.

It appears certain that these active substances originate in radioactive emanations in the atmosphere. The fact that they carry a positive charge is in keeping with this view, although the origin of this charge is not readily explained. The most conspicuous feature attending the transformations of the products of rapid change, arising from the break-up of a radioactive emanation, is the loss of the α particle. This is positively electrified, and we should expect that the residual atom must carry a negative charge, and would, therefore, be repelled rather than attracted by a negatively charged body. The electropositive sign of the residual atom is, however, abundantly proved. There is some evidence that the expulsion of the α particle from the transforming atoms may be attended, or followed, by the escape of slowly moving electrons. The intimate nature of the changes succeeding the expulsion of the helium atom is as yet unknown to us. There is reason to believe that even in the electric field the atom remains suspended for some moments after its gaseous properties have ceased to exist, during which its internal changes are progressing.[1]

The active deposit detected by Elster and Geitel has been obtained in many parts of the world, and at high altitudes on mountain tops. It has been found in rain-water and in snow. Considerable variations in its amount have, however, been observed. Elster and Geitel found that low barometric pressures were attended by greater quantities of the radioactive substances. Rutherford and Allen,[2] in Canada, perceived a variation according as the weather was clear and bright or otherwise, the clear weather being accompanied by larger amounts. Dull warm weather in summer was unfavourable to their presence. Eve[3] carried on observations for a year in Montreal, and found that temperature conditions appeared to be without effect; as much radioativity existed in the atmosphere when, in March, two or three feet of snow were on the ground,

[1] Rutherford, *Radioactivity*, p. 318.

[2] Rutherford and Allen, *Phil. Mag.*, Dec., 1902.

[3] Eve, *Phil. Mag.*, Oct., 1908.

as, in April, when the snow was gone. The approach of a deep cyclone caused increase; while anticyclonic conditions caused a diminution of the radioactivity.

Whence comes the active deposit? Several observers have found that the rate of decay of the activity of the deposit obtained on a negatively charged wire is about the same as that of the excited activity or active deposit of radium emanation. This evidence is strengthened by the fact that the air drawn from soils or found in caves also yields an active substance having the properties of this emanation; and in such localities we reasonably expect the presence of the emanation of radium. Again, it has been observed, in many cases, that well-waters bring to the surface dissolved radium emanation. At an early stage in these investigations, Sir J. J. Thomson showed that this was true of the Cambridge well-water. There is thus a considerable *à priori* probability that we should find this radioactive gas also in the atmosphere.

Eve[1] estimated the actual amount of radium emanation in the atmosphere by abstracting the active deposit contained in air enclosed in a large tank. Comparing the ionization produced by the abstracted deposit with the ionization due to a known amount of the active deposit of radium emanation, he concluded that in a cubic metre of air the amount of emanation which would be in equilibrium with 80×10^{-12} grams of radium must be present.

The most immediately important question respecting the subject of the ionization of the atmosphere

[1] Eve, *Phil. Mag.*, July, 1905.

is as to how far the radium detected everywhere in the earth's surface-materials is adequate to occasion the whole ionization, or whether some other source must be looked for. The evidence is conflicting. Eve[1] cites an experiment which would appear to show that the emanation of radium is the principal source of ionization. Using a very large iron tank, in which the normal rate of production of ions was so small as 9·6 per second per cubic centimetre, he withdrew the active deposit in the manner before referred to; that is by exposing a negatively electrified wire in the tank. The excited activity withdrawn upon the wire was estimated by its ionization. This experiment was then repeated in a smaller tank, using the emanation of radium. The ratio of the excited activity on the wire to the total ionization in the large tank was found to be 7/10ths of the same ratio found for the emanation of radium. From this it would appear that the larger part of the ionization in the air within the tank was due to the active deposit withdrawn from it. But it would seem that this experiment is not supported by actual estimates of the ionization to be expected from the observed amount of emanation in the air. We can easily calculate from known data the number of ions which would be produced per second from the quantity of emanation supplying the active deposit in a cubic metre of air (p. 19). Satterly,[2] who has made careful determinations of the amount of active deposit in the air at Cambridge, finds that the emanation is present to an amount representing on an average 100×10^{-12}

[1] Eve, *loc. cit.* [2] Satterly, *Phil. Mag.*, Oct., 1908.

grams of radium per cubic metre. Rutherford and Geiger have shown that the data respecting the ionization produced by an alpha particle of radium, and the numbers of such particles produced per second by a known amount of radium, require revision. Thus the number of ions created by a single α ray of radium is 100,000 (p. 15), and $3{\cdot}4 \times 10^{10}$ helium atoms are radiated per second from one gram of radium. Boltwood[1] has determined the relative activities, when in mutual equilibrium, of radium, emanation, and the substances entering into the active deposit. Taking the ionizing activity of uranium as unity, the values are :—

Radium, ...	0·45
Emanation, . . .	0·62
Radium A, . . .	0·54
„ C, . . .	0·91

When dealing with the effects of the emanation on the atmosphere, we are concerned with the last three elements only. We must conclude from the relative activities (as the same number of rays is emitted per second by each when they are in equilibrium) that if the α ray of radium creates 100,000 ions, the last three elements, for each α ray which they emit, create respectively 140,000, 120,000, and 200,000 ions. Thus in one second the emanation of one gram of radium in equilibrium with its rapidly changing transformation products, will give rise to $460{,}000 \times 3{\cdot}4 \times 10^{10}$ ions. Taking now the emanation in the atmosphere as representing as much as 100×10^{-12} grams of radium per cubic metre, and multiplying by this value, we conclude

[1] Boltwood, *Am. Journ. Sc.*, Ap., 1908, p. 297.

that the creation of 1·56 ion per cubic centimetre per second can be ascribed to the amount of radium emanation present. As we have already seen, this is in general insufficient to account for the whole ionization of the atmosphere.

It may here be mentioned that many recent measurements of the radioactive substances in the atmosphere have been effected by the direct abstraction of the emanation. Thus, Eve,[1] Ashman,[2] and Satterly[3] have used either the absorptive properties of charcoal, discovered by Rutherford, or the low temperatures obtained by the use of liquid air, to condense it directly from the air; in this way bringing into the electroscope the emanation from one hundred or two hundred litres of air. The emanation so extracted exhibits the same properties as the emanation of radium; and the amounts obtained confirm Eve's original estimate of its average quantity in the atmosphere. The variations from day to day are considerable. Thus Ashman finds the equivalent radium to vary from 45×10^{-12} to 200×10^{-12} grams per cubic metre. Eve[4] finds the variation as much as 1 to 7.

If it be assumed that the radium emanation found to exist in the atmosphere is derived from the rocks and soils, we are enabled to make some estimate of the percentage of the emanation evolved which escapes into the atmosphere. We must assume a limiting depth below which the emanation is completely conserved, or

[1] Eve, *Phil. Mag.*, Oct., 1908.

[2] Ashman, *Am. Journ. Sc.* xxvi., 119, 1908.

[3] Satterly, *loc. cit.* [4] Eve, *loc. cit.*

from which it cannot reach the surface in a sufficiently short time to be effective. This limiting depth must, in general, be close to the surface. Most soils extend downward but a few feet before they give place to the fine, compacted particles constituting the subsoil. Diffusion through this close-grained and often water-logged medium must be slow. Again, other soils are limited by rock, from which little or no emanation can escape. It would seem that an average depth of two metres would be a liberal estimate. Near the surface of this, the diffusion of emanation into the atmosphere will be rapid; but, as we go downward, the time taken to reach the surface must result in the decay of much of the emanation. We will assume, however, that the whole of the emanation escaping from the mineral particles succeeds in entering the atmosphere undiminished in quantity. In calculating the amount of emanation in the atmosphere, we may take—following Eve—a height of ten kilometres, within which it is permissible to apply the estimates of emanation made on plains and mountains.

If now we take the radioactivity of the soil as 8×10^{-12} grams of radium per cubic centimetre or 8×10^{-6} per cubic metre, and that of the atmosphere as 80×10^{-12} per cubic metre, we have

$$x \times 2 \times 8 \times 10^{-6} = 80 \times 10^{-12} \times 10^{4},$$

from which x, the radium in the soil supplying emanation to the atmosphere, is found to be 5 per cent. Experiments directed to the quantitative measurement of the output of emanation of various soils might enable us to test the validity of the theory which ascribes the emanation in the atmosphere to the radium in the soils. That some emanation originates in the soils is certain; for, in fact, it has been observed *in situ.*

The ionization over the ocean has been made the subject of investigation by A. Boltzmann and by Eve.[1] The observations in both cases were made upon the North Atlantic; and the results were concordant in showing that there is not much difference in the number of ions in air over the ocean and over the land. Boltzmann's average is 1150 + ions and 800 − ions; numbers which do not differ notably from results obtained in Lapland or Germany. Eve quotes from G. C. Simpson, an average for 1903–1904, obtained in Lapland, of 1120 + and 972 − ions.

Eve made observations from Montreal to the Isle of Man. The readings at Montreal were unusually low; and in the Gulf of St. Lawrence the rise in the number of ions was rapid and continuous. Some of the highest readings were obtained in mid-ocean; but they did not exceed the ordinary readings observed upon the land.

Now under the conditions of Eve's observations, there would be difficulty in explaining the presence of emanation over the sea, if the sea was not itself a source of radioactive supply. It must be remembered that the perishable nature of the emanation differentiates it from an ordinary gas. The general uniformity in composition of the atmosphere is easily explained in its mobility and circulation. But the emanation of radium decays to half its amount in ninety

[1] Eve, *Phil. Mag.*, February, 1907.

hours. It cannot take long journeys unless borne by strong winds. In anticyclonic conditions, its presence a thousand miles from land, to the same amount as over the surface from which it is derived, would be inexplicable. We must therefore consider the ocean a source of radium emanation. It is worthy of notice that the transit from the land to the central seas of the emanations of thorium or actinium is an impossibility; the gaseous state in these cases practically disappearing in a few minutes.

It is true that the average radioactivity of sea-water is only about the $\frac{1}{500}$ part of the radioactivity of the rocks (p. 46). But, on the other hand, the radium is *in solution*, and may be shown by experiment to yield its emanation freely to the air. And again, there is perpetual renewal of the surface supply; so that if by diffusion into the atmosphere the emanation falls below the equilibrium amount, water rising from beneath must continually maintain the maximal conditions at the surface. It is also well known that bubbling air through radioactive liquids is a highly effective—although not perfect—means of extracting the emanation. Every bubble rising to the surface therefore brings emanation into the atmosphere. In rough and stormy weather we should expect that the supply of radioactive gas to the air must be very considerable. Although the specific radioactivity of sea-water is so inferior to that of the rocks, the dissolved state of the radium, as well as the mobility and limitless amount of the medium in which it is contained, may render the radium content of the ocean a sufficient explanation of the ionization observed. Estimates of the emanation over the central parts of the ocean, and contained in the surface waters, would be of much interest.

We shall now consider another source of ionization arising from the presence of radioactive elements at the surface of the earth—penetrating γ and β rays. So far as these rays are referable to the known amount of radium emanation contained in the air, their effects must be insignificant—a small fraction only of the ionization due to the α rays. It is otherwise, however, when we come to consider these radiations as emitted from the radioactive elements contained in the rocks and soils.

That a very penetrating radiation affects gases near the surface of the earth seems proved by the experiments of M'Lennan[1] and Rutherford and Cooke[2] on the screening action of thick metal walls placed around closed vessels in which the rate of production of ions is under observation. Thus Cooke found that in a brass electroscope the rate of production of ions was 13·6 per cubic centimetre per second. But on encircling the vessel with walls of lead 5 centimetres thick, this number was reduced to 9·1. It appears from this, clearly, that not less than 4·5 ions per second were being produced by some penetrating radiation before the lead plates were in place.

An interesting calculation of Eve's appears to show that this penetrating radiation may be ascribed, mainly, to the radium in the rocks and soils. As the results of McClelland's investigations on the penetration of γ

[1] M'Lennan, *Phys. Rev.*, 4, 1903.

[2] Rutherford and Cooke, *Amer. Phys. Soc.*, Dec. 1902.

rays,[1] Eve finds that these rays will be cut down to half their amount after traversing 7·5 centimetres of average rock, and to about 1 per cent. after passing through 50 centimetres. There is then a strictly limited depth from which the penetrating radiation must be supposed to proceed. The experiment of Cooke indicates that at the surface of the earth 4·5 ions per cubic centimetre per second arise from these rays in air contained in a brass vessel. This number would be increased if the electroscope was entirely under ground. The number of ions generated should be doubled: in other words, there should be 9 generated per second per cubic centimetre. As it has been found that the ionization produced within a vessel, by penetrating radiation, varies with the nature of the material of which the vessel is made, a part of the effects in Cooke's experiment must be ascribed to a secondary radiation proceeding from the brass walls of the vessel. Eve, however, connects the ionization observed, with the amount of radium giving rise to the penetrating rays, by an independent experiment on the ionization produced in such a vessel by a known quantity of radium. This gives us the means of estimating the amount of radium in the rocks which would be required to give rise to 9 ions per cubic centimetre per second in the brass vessel when buried in the ground; or, conversely, knowing the radioactivity of the rocks, we might calculate the ionization in the electroscope. Eve estimates the radium, and finds that the radioactivity of the rocks required to produce the ionization observed by

[1] McClelland, *Phil. Mag.*, July, 1904.

Cooke, must be 10×10^{-12} grams of radium per cubic centimetre of rock. This result is in satisfactory agreement with those recorded in Chapter III for many varieties of rock, but is rather high for the leached-out materials of the soil; and points, perhaps, to the presence of other radioactive substances in the surface-materials.

Eve concludes that the ionization in the atmosphere can, in fact, be accounted for by the joint effects of the radium emanation actually present in it and the penetrating rays coming from the rocks and soils. The penetrating radiation from radioactive materials at the immediate surface would be able to traverse the atmosphere to a distance of 1000 metres before 99 per cent. of it was absorbed. At 157 metres it would be absorbed to the extent of one-half.

More recently C. S. Wright,[1] experimenting on the ionization of air within a lead electroscope, as observed when placed over ordinary soil, and again when over the radium-free water of Lake Ontario, obtained results which would seem to show that from 2·6 to 4·6 ions are produced per second per cubic centimetre in the electroscope by penetrating radiation; and concludes, from M'Lennan's experiments on the secondary radiation in lead vessels, that only one third the number of ions produced in the electroscope would actually be created in free air.

It would appear from the experimental work which we have now reviewed that, on the whole, the balance of evidence seems opposed to the view that the ioniz-

[1] Wright, *Phil. Mag.*, Feb., 1909, p. 295.

ation of the lower atmosphere is everywhere entirely due to the effects of radium emanation in the air and of radium in the rocks. This conclusion is strengthened by the positive evidence, which lately has been growing in force, for the existence in the atmosphere of radioactive substances not of the uranium series, but referable to the parent radioactive element thorium.

It has been found by several observers that the active deposit condensed on a negatively charged wire exposed for a sufficiently prolonged time to the atmosphere is, for a considerable part, that arising from the break-up of the emanation of thorium. Thus Bumstead referred to thorium from 3 to 15 per cent. of the whole amount of deposit obtained from air at New Haven, Conn.[1] Underground air from the same locality was found by Dadourian[2] to contain quantities of thorium emanation. Similar results have been obtained by G. A. Blanc at Rome,[3] who estimated that from 50 to 70 per cent. of the whole activity of the deposit condensed from the atmosphere was due to thorium. The escape of an emanation, which exists but for a few minutes before its amount becomes negligibly small, from any depth in the soil, seems impossible; so that the recorded observations appear to indicate a very considerable content of thorium in the rocks. Blanc[4] believes that its active deposit moves about attached to dust particles.

[1] Bumstead, *Am. Journ. Sc.*, July, 1904.

[2] Dadourian, *Am. Journ. Sc.*, Jan., 1905.

[3] Blanc, *Phil. Mag.*, March, 1907.

[4] Blanc, *Phys. Zeitsch.*, ix., 255, 1908

W. Wilson[1] examining the deposit condensed on a negatively charged wire exposed at Manchester, infers from the amount of the deposit of thorium collected, that about seven times as much thorium as uranium must exist in the soils. As the rate of transformation of thorium is, according to Bragg,[2] one-fifth that of uranium, and as there are six helium-producing products to eight in the case of the uranium series, the energy would be approximately alike in each case. If this be generally true, the geological importance of thorium would be equal to that of uranium.

The problem of the continued sustenance of the ionization of the atmosphere cannot be said to be solved. But there can be no doubt of the important part played by radioactivity; and herein we find another manifestation of the influence of the radioactive elements on geological processes.

We might discuss other effects upon the atmosphere not improbably connected with radioactivity. The time for doing so has, however, hardly come. Some of its rarer constituents may be ascribable to radioactive sources. Helium is very probably so derived. We have seen how its energetic flight from the atom is one source of atmospheric ionization. The trace of hydrogen present may in part arise from the radioactive decomposition of water, which, in a minute degree, must be progressing not only where moisture is present in the soils, but throughout the ocean. In the case of these two gases there is the additional

[1] Wilson, *Phil. Mag.*, Feb., 1909, p. 321.

[2] Bragg, *Phil. Mag.*, June, 1906, p. 754.

ground for assuming their radioactive origin that, on the theory put forward by Stoney, there must be continual loss of these lighter gaseous molecules from the earth's atmosphere; and hence we infer that some steady source of supply must exist. It might be expected that traces of chlorine should reach the atmosphere through the agency of the radio-chemistry of the ocean and its sediments; Cameron and Ramsay[1] having shown that hydrochloric acid is decomposed by radium emanation. A computation of the amount which might be produced in this way seems, however, to indicate that it must be very minute, notwithstanding the enormous reservoir within which the processes are at work.

Sir William Ramsay's recent correction[2] on his estimate of the helium in the atmosphere (p. 63), increasing it to 4 parts in 10,000, involves the existence of 16×10^{14} cubic metres of atmospheric helium, taking the height of the homogeneous atmosphere as 8 kilometres. So great a quantity, if of radioactive origin, cannot have been generated during geological time. For it is easy to show that if all the heat leaving the earth be referable to radioactive energy, the equivalent evolution of helium is 36×10^6 cubic metres per annum; and of this only a small fraction can enter the atmosphere. If the revised figures are applicable to the atmosphere generally, we may, perhaps, have here an indication of pre-geological time, and evidence sustaining our highest estimates of radioactive matter near the surface of the earth; in other words, of the radioactive origin of terrestrial heat.

[1] Cameron and Ramsay, *Chem. Soc. Journ.*, June, 1908, p. 984.

[2] Ramsay, *Proc. R. S.*, 80 A, 599

CHAPTER XI.

URANIUM AND THE AGE OF THE EARTH.

PRINCIPLES OF THE RADIOACTIVE METHOD OF ESTIMATING THE GEOLOGICAL AGE—HELIUM AND THE ALPHA RAY—RATE OF COLLECTION OF HELIUM—RUTHERFORD'S ESTIMATE OF AGE OF MINERALS—STRUTT'S EXPLORATIONS FOR HELIUM—HIS WORK ON GEOLOGICAL TIME—CRITICISM OF THE RADIOACTIVE METHOD—MIGRATIONS OF RADIOACTIVE SUBSTANCES—CHANGES IN PHOSPHATIC SUBSTANCES—HELIUM AND THORIUM—ACCUMULATION OF LEAD AND AGE OF MINERALS—BOLTWOOD'S WORK—CONTRADICTION BETWEEN RADIOACTIVE METHOD AND METHODS BASED ON DENUDATION—THE METHOD BY SEDIMENTATION—SOURCES OF ERROR—THE METHOD BY SOLVENT DENUDATION—CONSIDERATION OF THE CORRECTIONS—LEGITIMACY OF UNIFORMITARIAN ASSUMPTIONS.

SUNK in depths of time which, mentally, are unfathomable, the earth's beginning has yet ever been the subject of philosophic questioning or scientific exploration. The deciphering by geologists of a great sequential history in the piled-up debris of the mountains, while impressively suggesting the vastness of the duration involved, has left unsatisfied our craving for a knowledge of the interval since the records began. The same overwhelming impression, the same fascination of the unknown, are experienced, whether we endeavour to review those activities which slowly laid the foundations of mountains, uplifted and again destroyed them; or whether we contemplate the biological past which stands to our brief length of days as the integral life to the life of the individual. The primary question of historical geology is the age which has encompassed the collective events since the beginning.

The age of the earth may indeed claim an importance outside of geological science. For is it not the clue to the chronological scale of our solar system and—for want of a better—even of our universe? Distinct from the age of the earth, the geological age, or period since denudation began and life became possible, if of lesser cosmic importance, is still a minor limit of priceless value. If geological time had been—as once was believed—almost eternal in duration, then must our conception of solar chronology be enlarged accordingly. If, on the other hand, Kelvin's twenty millions of years could be established, then must solar events be, comparatively speaking, short-lived, and the whole cosmos may have entered on its present phase of evolution within a period correspondingly recent.

It is, in a word, the primary chronological fact of inorganic evolution, as it is of organic evolution; and to it our fundamental ideas of duration must be referred, as the depths of space are referred to the distance separating earth and sun.

The discoveries of radioactive science early suggested the possibility of approaching the problem of the geological age in a new way. Uranium is constantly breaking down into an element of less atomic weight; and this transformation progresses alike within the hardest rocks and in the waters of the ocean. The derived element changes to one of yet lower atomic weight, and so this material degradation proceeds until a stable substance is reached at which the change ceases; or ceases for the present time. Now, what becomes of the lost mass? We have no reason to believe that any part of it is destroyed. Indeed, we know that it is not destroyed; we know that it is radiated in the alpha particles, the energy of which has been so important a factor in geological dynamics. The process of transformation is thus attended with the rejection of portions of the atom at every change of atomic weight. The discarded particles are physically alike in every case, so far as we know; and, as Rutherford long ago predicted, are in themselves a recognized kind of matter—helium.

This substance, radiated in an ionic or electrified state, accumulates amidst the changing matter which gives rise to it, unless there be some avenue of escape. In the case of the uranium family the number of alpha rays or helium atoms, which the entire sequence of changes involves, is fairly well known. If we know the rate at which these changes progress—and this knowledge is also ours to a sufficient degree of accuracy—and in any particular case determine both the amounts of radioactive substances present and the quantity of accumulated helium, we are evidently in possession of all the data required to determine the period during which the changes have been going on.

The identity of the alpha ray with helium is a fundamental point in this procedure. The evidence on which this identification is founded may be taken as conclusive. Rutherford and Geiger[1] have shown

[1] *Proc. R.S.*, 81 A, p. 162.

by a very beautiful investigation that the charge on a single alpha particle is about twice the charge carried by an ionized atom of hydrogen; and hence by a knowledge of the ratio of the charge to the mass, the mass is found to be 3·84 times that of the hydrogen atom; in other words, the atomic weight is 3·84. Now, quite independently it is known that the atomic weight of helium is 3·96. The agreement is remarkable, considering the delicate nature of the measurements involved.

In a second investigation[1] the purified emanation of radium is introduced into a glass tube so thin in the walls that the alpha rays pass through, while the emanation and its transformation products remain enclosed. This tube is placed within a larger one rendered highly vacuous. A spectroscopic examination of the residual gas in this outer tube at first shows no helium, but as time progresses the helium spectrum appears. This shows that the alpha rays, passing through the walls of the inner tube and trapped within the outer one, are helium. Here we are secure from the possibility that the helium known to come into existence in the transformations of radium and its descendants, is some resultant of the changes other than the alpha ray itself. Finally, if any further proof was required, Dewar[2] has measured the rate of production of helium from radium, and finds that this is in close agreement with that which Rutherford had predicted on a count of the emitted alpha particles: that is 0·37 cubic millimetres per gram of radium per day.

[1] Rutherford and Royd, *Proc. Manchester Lit. and Phil. Soc.*, 53, p. 1.

[2] Sir James Dewar, *Proc. R. S.*, 81 A, p. 280.

The diagram of the radioactive elements given in Chapter I shows that in the direct line of descent, the atom of uranium emits two alpha particles; and ionium, radium, emanation, radium *A*, radium *C*, and polonium, each emit one particle. When therefore these substances exist together in equilibrium, as they would in an old mineral, every atom of the parent substance which transforms involves, on the simplest assumption, the production of a total of 8 atoms of helium. Additional to this source of helium, the actinium branch supplies a small quota. But this must be a comparatively unimportant addition, the activity of actinium in uranium-bearing minerals being only a feeble part of the whole.[1]

At the time when Rutherford[2] first applied this method to determine the age of radioactive minerals, the constants were not so well known as they are now. The original calculations are accordingly in need of correction. He cites the case of a specimen of fergusonite which Ramsay and Travers found to contain 1·81 cubic centimetres of helium per gram, and about 7 per cent. of uranium. We now know that the ratio of uranium to radium in minerals must be closely as 1 is to $3{\cdot}4 \times 10^{-7}$. The quantity of radium present must, therefore, be $23{\cdot}8 \times 10^{-9}$ grams per gram. Now, $3{\cdot}4 \times 10^{10}$ alpha particles are expelled per second per gram of radium,[3] or if we count up the eight rays which appear involved in the theory of radioactive change,

[1] Boltwood, *American Journ. Sc.*, April, 1908, p. 291.

[2] Rutherford, *Phil. Mag.*, Oct., 1906, p. 368.

[3] Rutherford and Geiger, *loc. cit.*

for each atom of radium breaking up, $27 \cdot 2 \times 10^{10}$ alpha particles are emitted; and this is the number expelled per second per gram of radium. Taking now $2 \cdot 72 \times 10^{19}$ molecules as the number of molecules in one cubic centimetre of gas at standard pressure and temperature, we would have 10^{-8} cubic centimetres of helium evolved per second in a hypothetical mineral containing one gram of radium. This is equal to 316 cubic millimetres per year. But in the fergusonite the amount of radium is only $23 \cdot 8 \times 10^{-9}$, and therefore the annual evolution of helium is $0 \cdot 75 \times 10^{-8}$ cubic centimetres per gram. We can from this calculate the age. The helium in the mineral was found to be, as stated, 1·81 cubic centimetres per gram: dividing this number by the amount of helium evolved per year in one gram of this mineral, we find the duration required for the accumulation of the helium was 241 millions of years. Nothing is said about the probable geological position of this mineral. It may possibly date from a time before our epoch of denudative history begins. Rutherford, however, suggests that the age of the geological strata might also be investigated in this manner.

This calculation may be taken as typical of the steps involved. It will be seen that we have to measure not only the helium, but also the uranium or the radium. If we take the radium—and of necessity we must in general do so—we may proceed to find the age on the rate of production of helium which the presence of the observed amount of radium involves. Or we may make a similar calculation, working directly from a knowledge of the uranium present.

According to the experimental value found by Sir James Dewar for the rate of evolution of helium, the quantity is 135 cubic millimetres per year per gram of radium. This, however, is only for radium in equilibrium with its products of rapid change. It therefore only includes four out of the eight alpha particles which in an old mineral containing the complete series would be emitted. Doubling his figure, we find the constant is 270 cubic millimetres per annum. This would add about seventeen per cent. to the age as determined on the data cited above.

Quite recently Soddy[1] has measured the helium evolved by uranium itself, and obtained quantities which suggest that Boltwood's view, that two alpha rays are evolved from each transforming atom of uranium, is incorrect. This observation, however, does not, probably, point to a lessened rate of production of helium, but rather to the existence of some unrecognized element in the uranium series. The inferred rate of production of helium, in the case of uranium and some other of the radioactive elements, rests upon measurements of the relative ionizing activity; the helium out-put in the case of radium and its products of rapid change being, as we have seen, directly determined.

[1] Soddy, *Nature*, Dec. 3, 1908, p. 129.

In a series of papers appearing in the *Proceedings of the Royal Society* within the last couple of years, Strutt has gone into the question of the occurrence of helium as a widespread constituent of the earth's crust, more completely than any of his predecessors. Subsequent to Ramsay's discovery of helium in cleveite and other minerals, Lord Rayleigh[1] in 1896 showed that helium was a constituent of the hot springs at Bath; and Moureu[2] found that, in hot springs occurring in France, helium was a common ingredient; and this observation has since been confirmed for thermal springs in various parts of the world. Natural gases obtained from the earth have also been shown to contain helium as a general constituent. Thus Cady and McFarland,[3] in forty-one sources of natural gas in America, found in all cases helium present from traces up to as much as 1·84 per cent.

It would seem from Strutt's observations that we need look no further than the radioactivity of common rocks and minerals for these supplies of helium. In a large number of rare and common minerals, in rocks both igneous and sedimentary, helium occurs; and most generally in quantities which might be referred to the uranium-radium elements present.[4] Exceptions to this have, however, been observed. For instance,

[1] Lord Rayleigh, *Proc. R. S.*, 59, p. 198.

[2] *Compte Rendu*, 1906, v. 152, p. 1155.

[3] *Annual Report of Chemical Soc. on Radioactivity*, by F. Soddy, 1907, p. 323.

[4] Strutt, *Proc. R. S.*, 80 A, p. 572.

a fluorite wherein a large quantity of helium can only be ascribed to thorium,[1] and beryl, wherein the helium occurs without any adequate quantity of radioactive elements to account for its presence. Again, Strutt[2] finds reason to believe that in the Stassfurt salt deposits, helium is liberated by the element potassium; the radioactive properties of which have lately been shown by the researches of Campbell and Wood.[3]

Throughout the results obtained by Strutt there is a very wide variation in the ratio of the quantity of helium to the quantity of uranium oxide as calculated from the radium. This wide variation seems impossible to account for on the score of the age of the substance; indeed, as Strutt points out, we find in the Cornish granites a considerably smaller ratio than obtains in the minerals of the veins traversing it, and which are necessarily younger in age.[4]

More recently Strutt has approached the problem of the geological age directly; taking as materials for examination substances which must be contemporaneous in age with the strata in which they are found. More especially he makes use of phosphatic nodules and phosphatized bones, which he finds to contain an unusually large amount of radinm; sometimes up to fifty times that found in common rocks. The following table summarizes his results.[5]

[1] Strutt, *Proc. R. S.*, 80 A, p. 56.

[2] *Ibid.*, 81 A, p. 278.

[3] Campbell and Wood, *Camb. Phil. Soc. Proc.*, vol. 19, p. 15.

[4] Strutt, *Proc. R. S.*, 80 A, p. 582.

[5] *Ibid.*, 81 A, p. 272.

Material.	Locality.	Geological Horizon.	Years.
Phosphatized sharks' teeth, .	Florida, ...	Pliocene, ...	$77 + 10^3$
,, Cetacean bones,	Felixstowe, ...	Pliocene Red Crag, . .	112 ,,
Phosphatic nodules, . .	,, ...	,, ,, . . .	225 ,,
,, ,, . .	Cambridge, ...	Upper Greensand, . .	$3{\cdot}08 + 10^6$
,, ,, . .	Potton, Bedfordshire, . .	Lower ,, . . .	3·95 ,,
Phosphatized Saurian bones,	Ely, ...	Kimmeridge Clay, . . .	$>$1·21 ,,
Phosphatic nodules, . .	Knapwell, Cambs., . .	Base of Kimmeridge Clay, .	$>$1·03 ,,
Phosphatized Saurian bones,	Whittlesea, . , . .	Oxford Clay, ...	$>$6·11 ,,
Phosphatic bone fragments, .	Lyme Regis, ...	Rhaetic bone bed, . . .	$>$1·12 ,,
Hæmatite, ...	Frizington, Cumberland, .	Above Carboniferous limestone,	141 ,,
Phosphatic nodules, . .	Near Bala, ...	Bala beds, ...	51·9 ,,
,, limestone, . .	Chirbury, Shropshire, . .	Llandeilo limestone, . .	77·9 ,,
,, nodules, . .	Cailleach Head, Loch Broom,	Torridon Sandstone, . .	9·25 ,,

The last column has been deduced on data similar to those already discussed in relation to the age of a specimen of fergusonite, but calculating, for convenience in dealing with the data used by Strutt, the annual production of helium from one gram of uranium oxide ($U_3 O_8$). This constant is found to be $9{\cdot}13 \times 10^{-8}$ cubic centimetres.

The variations of apparent age, irrespective of the order which must be assigned from the geological position, are very considerable. Strutt considers this is best explained as due to escape of helium, regarding the deduced age as a minor limit.

We will doubtless soon have more to go upon in judging of the possibilities of this method according as the errors affecting it and causing the variations are investigated. We cannot, however, discuss its importance without indicating the directions in which error may lessen its value.

The assumptions involved are that at the time of formation of the mineral substance a certain amount of uranium was introduced, which continued in its present position, only diminishing by transformation; and that all its products of change also remained, neither increased nor diminished, preserved in a similar manner within the substance. The gravest objection which can be brought against the method is, of course, the risk that these conditions are not likely all to be fulfilled in any one case, and that we are in ignorance of the fact whether they are fulfilled or not. Only in the event of arriving at concordant results under different circumstances of material and preservation would there be any assurance that some one or other of the conditions had not broken down during the long past.

These doubts arise in view of the known potency of underground waters in producing changes, either of subtractive, additive, or metasomatic character, throughout even hundreds of feet of rock. When along with this fact the geological history of even the most undisturbed strata is considered—the changes of temperature and pressure which must have assailed them—and recollecting that many millions of years are involved in the case of even comparatively modern rocks—the chances of finding helium, uranium, and radium, and all the chain of elements neither increased nor diminished must be remote. The retentive and stable nature of the material selected for investigation will be in this connexion an element of importance.

On the other hand might be urged the fact that in uranium ores we find a satisfactory degree of uniformity in the ratio of uranium to radium. It would appear from Boltwood's observations, however, that the constancy of the helium ratio is less in evidence, and, indeed, widely departed from.[1] Again, it should be borne in mind as generally true, that great losses in the past, of uranium or radium, or even gains of these substances, might occur; but if the conditions of change had been so far mitigated, or had so far exhausted themselves for the last two thousand or three thousand years, as to suffer equilibrium to be established, we would remain in ignorance of the losses

[1] Boltwood, *Am. Journ. Science*, Feb., 1907, p. 77.

or gains of the millions of antecedent years. The possibilities are indicated in the formation of secondary ores such as gummite, autunite, uranophane, and secondary pitchblende; all fairly common in association with uranium minerals, and many found deposited in comparatively recent geological times. And with reference to even the harder crystalline minerals, it must be noted that Jaquerod and Perrot have shown that helium permeates quartz at temperatures above 220° C, and with a velocity which increases with the temperature.[1] This would seem to render untrustworthy, readings obtained from metamorphic rocks, or rocks which had taken part in great earth movements.

The removal and redeposition of radium are not merely hypothetical. Danne records a remarkable case.[2] The radium was found in deposits of phosphate of lead (pyromorphite), in clays containing lead, and in pegmatite, but most abundantly in the phosphate. No trace of uranium was present. To reach its present position in the pyromorphite, the radium must penetrate veins of quartz and felspar—moisture due to neighbouring springs being the probable carrier. The quantity of radium varied considerably, and Danne believes that about a hundredth part of a gram per tonne must be present. This is, of course, a very large amount. The source appears indicated in the occurrence of autunite in this region, and it is stated as probable that other occurrences of uranium exist. The storage of radium is in this case traceable, very

[1] *Nature*, Oct. 22, 1908, p. 635.

[2] See Rutherford's *Radioactivity*, p. 465.

probably, to carriage by water in past ages. The concentration in the phosphate is remarkable.

There is some reason to fear that phosphatic substances, as repositories of radioactive elements and their products of change, may prove unreliable. In the first place, it is evident that at some time in the history of the phosphatic nodules there must have been an accretion or absorption of uranium or radium, or of both. For, in fact, these bodies are of organic origin, and probably would not originally have possessed their radioactive materials. When did this accretion occur? The most probable answer is that it was effected upon the sea-floor. But are we justified in assuming that it did not continue for a very long time after? Knowing from the facts of their genesis that accretion or absorption of radioactive elements occurred at one time in their career, and also knowing that underground waters are not infrequently charged with small quantities of radium, and often with helium, why should we limit the absorption to the geological period in which the nodules are found? If this reasoning is correct, there is possibility of results either unduly great or unduly little, according to very complex conditions in the long past.

The history of these phosphatic bodies shows that they have had a chequered career. They originated as alteration products of organic remains thrown down on the floor of the ocean, where they probably gathered radium or uranium, and, indeed, almost certainly both, the surroundings being highly radioactive. Their sojourn under these conditions may have been thousands of years: possibly in some cases much longer. If during this period radium was directly absorbed, a positive source of error in the accumulated helium is incurred to start with.

The phosphatization of these bodies is by no means completed in their original surroundings. A further concentration is effected after the sediments have been elevated into land surfaces; when, by leaching with meteoric or subterranean waters, the excess of calcium carbonate is washed away.[1]

The phosphorites of Tennessee are cited by Clarke as "an excellent illustration of the several processes, chemical and mechanical, which have taken part in their formation." These were partly Ordovician and partly Devonian, and as phosphatic limestones deriving their phosphates from bones, shells, and teeth, were first laid down in a shallow sea. There was then leaching of the limestones, which removed carbonates, leaving a mixture of phosphates, clay, and iron hydroxide. The soil thus formed was again concentrated by mechanical washing, the moving waters carrying away the clay and leaving the heavy phosphatic nodules. Some of the phosphates were actually dissolved by percolating waters and precipitated as secondary deposits in the underlying limestones. All this indicates that in these substances we deal with bodies specially liable to alteration. That these alterations may take the form of the addition of substances to the phosphate, appears from the following facts:—

"In certain Cretaceous sandstones of Russia calcium

[1] Clarke, *The Data of Geochemistry*, p. 448.

phosphate occurs as a cement for the sand grains, and also in the form of fossil-bones and fossil wood. The wood has been completely replaced by the phosphate. Although bone is itself largely composed of calcium phosphate, fossil bones are not identical chemically with recent bones. The fossils show an enrichment in calcium carbonate, iron oxide, and fluorine, as A. Carnot has shown, and especially in fluorine. Modern bones from various animals were found by Carnot to contain a minimum proportion of fluorine; Tertiary bones were much richer; Triassic and Cretaceous bones still more so; and in bones from Devonian and Silurian formations the ratio of fluoride to phosphate was nearly that of apatite. This progressive enrichment in fluorine Carnot attributes to the agency of percolating waters, carrying small quantities of fluorides in solution."[1] Bischof knew of the absorption by fossil bones of the element fluorine, and suggests a chemical explanation.[2]

In short, here are cumulative actions of just the kind we previously referred to as possibly resulting in the accretion of radioactive materials, unless we suppose the fluorine generated within the phosphate. The conclusion seems to be that a steady, mutual substantiation must be required of results which involve, to so great a degree, unaltered relations having existed between the mineral and its surroundings; and the further back we seek to extend our explorations, the more desirable is the assurance arising from mutual

[1] Clarke, *loc. cit.*, p. 447.

[2] Bischof, *Chemical and Physical Geology*, ii., p. 33.

support. That greater security might sometimes be gained by direct determination of the uranium is suggested by the presumptive evidence in favour of radium being captured by phosphates without the simultaneous storage of the parent element.

These objections unfortunately apply in some degree to any substance of secondary origin. Thus if we have to deal with a mineral substance probably derived by metasomatic change of a pre-existing rock, we have of necessity to ask if it is certain that the waters concerned in the genesis of the mineral brought in the uranium and radium, and that these elements were not introduced subsequently. And again, may not the latter element have taken up its position in quantities beyond the true equilibrium amount? Here again we seem entitled to ask for the corroboration of results, before we conclude that these sources of error are absent. In a word, when we examine into the claims of the radioactive method, we find that, in a certain sense, we postulate a degree of uniformity between present and past conditions which might give pause to the most advanced uniformitarian geologist. It has, however, this advantage over any other known method of approaching the problem of the earth's age, that each experiment is, even if uncertain, an independent estimate, and that we may make as many of them as we please. Thus it may result that out of a large number we may haply find a degree of coincidence sufficing to confer complete credibility on the evidence.

The production of helium from thorium, and the possibility of helium so produced making an appreciable part of the whole amount, have been carefully considered by Strutt.[1] He has brought forward much evidence to show that this may indeed be the case. It may be added that the fact that the alpha ray is a helium atom appears to set at rest the sometime disputed question whether thorium could indeed give rise to helium. Direct evidence of the genesis of helium by thorium has also been obtained by Soddy.[2] Obviously if this source of helium is of moment, its neglect in any particular experiment is in the direction of producing an error of excess in the age.

The idea that lead might be the ultimate disintegration product of uranium has been made the basis of an attack on the age problem. In a paper of great interest Boltwood[3] records that in unaltered primary minerals from the same locality, and presumably of the same age, the amount of lead was proportional to the amount of uranium present; and, further, that the amount of lead relative to the uranium was greatest in minerals from those localities which were of the greatest geological age. This, it will be admitted, is much in favour of the view that lead is indeed a disintegration product of uranium.

On the foregoing basis Boltwood arrives at the age of a number of uranium-bearing minerals. His figures, when corrected according to more recent data, respecting the constants involved (the proportion of uranium

[1] Strutt, *Proc. R. S.*, A 76, pp. 101, 312.

[2] Soddy, *Phil. Mag.*, October, 1908.

[3] Boltwood, *Am. Journ. Science*, February, 1907, p. 77.

to radium, and the rate of decay of radium), range from 246 to 1320 millions of years. The geological positions of these minerals are not cited; the oldest, from Norway and Ceylon, are presumably from primitive rocks.

It is not improbable that the radioactive method of determining the age will receive a useful development in the procedure adopted in this research. Although the introduction of radium, etc., unattended with uranium or the introduction of radioactive materials at dates subsequent to the formation of the substance, would give rise to larger errors than in the method based on the helium ratio, on the other hand, there would probably be more security against error from the loss of the disintegration product chosen for measurement.

The apparent tendency of Strutt's results is to assign ages to the older sedimentary deposits greater than would be deduced by the legitimate use of any other method of estimating the age of the earth with which we are acquainted. Thus Strutt selects the following results as showing the bearing of his measurements, regarding them as *minor limits*:—

	Years.
Phosphatic nodules of the Crag, . .	225,000
,, ,, Upper Greensand,	3,080,000
,, ,, Lower ,,	3,950,000
Hæmatite overlying Carboniferous Limestone	141,000,000

If the last of these estimates is correct, there can be

little doubt that the Huronian must date back some 400 millions of years, and the "geological age" would probably be somewhat greater still. It is, therefore, of importance to consider whether those existing methods, which have hitherto been interpreted as indicating from 50 millions to 100 millions of years as the period since denudation began, can be so far astray.

There exist, in fact, but two methods directly dealing with the denudation period: that which is based on the records of detrital denudation, and that which is based on the records of solvent denudation.

The first method seeks to find the period since sedimentation began by taking as numerator the maximum total depth of sedimentary deposits, and as denominator the rate of accumulation. The quotient is the period sought.

It must be admitted that grave difficulties attend the determination of both the numerator and denominator of this fraction:—

(*a*) What is the maximum thickness of the sediments? We are first met with the difficulty of not knowing where to begin. The more ancient sediments are apparently to be found in those complexes of igneous and sedimentary materials wherein the differentiation of the two classes is all but, or quite, impossible. About the origin of such rocks we find the most experienced observers differing. Again, even when the evidence is adjudged decisive as to an igneous origin, this may really denote no more than the complete igneous change of pre-existing sediments.

(*b*) A fine-grained deposit will take a far longer time to collect than an equal thickness of coarser materials. A whole formation will include layers of many different sorts. The time-values of these are very various; some are of slow, some of rapid deposition, suggesting that the area of maximum rate of deposition was not fixed in position. In seeking the denominator (the rate of accumulation) this must be remembered or serious error will be incurred.

(*c*) The unconformities often represent great lapses of time, which not only may be unrepresented, but may even have resulted in the removal of earlier deposits. As an instance, consider the almost complete removal of the mountain-building sediments of the Alps. This error may or may not be made good in the measurements at our disposal. Certain deposits are very perishable. The Chalk is an instance. This substance is not only perishable, but it is, relatively, a slow-forming deposit. In a few million years, if exposed so long to denudation, all records of such formations might be removed. Thus the time record would be falsified.

To these sources of error might be added such as may arise by the conditions being homotaxial rather than contemporaneous. For if Huxley's case have a semblance of the truth, and it be possible that the Devonian period in England was overlapped by the Silurian in America, and in turn overlapped the Carboniferous in Africa, we evidently are not entitled to sum the several depths of deposit in these countries as successive time intervals. This latter objection has hardly more than academic interest at the present moment. The radioactive method of investigating the ages of the strata may yet put Huxley's case to the test. Finally the practical difficulty of measuring the thickness of strata has to be faced. The difficulty of the task is shown in the diversity of opinion as to many of the recorded measurements.

Supposing we have satisfactorily determined the numerator, we are by no means done with our troubles. We have to find the rate at which the rivers are carrying detrital sediments to the ocean. This involves measurements of the silt-load of rivers, at various seasons. The seasonal variation is great, because the same cause which raises the river to flood-level increases the supply of suspended matter. The burden carried may be five-fold as much at one time as at another time.[1]

We have also to find the rate at which this silt grows in depth when it comes to rest in the ocean. And here is perhaps the most serious source of error, for not only is the catchment area difficult to define, but the troublesome question remains as to whether we should reckon the silt as laid down over the whole area, and take this mean rate of deposition, or, having found this average rate of accumulation, proceed to increase it in order to arrive at the maximum rate, having in view the fact that we have taken the sedimentary

[1] See table in Russell's *River Development*, p. 72.

records at their maximum thickness. The final result will vary considerably according to which procedure we adopt. Thus while Geikie[1] is satisfied that the whole geological column may have taken about 100 million years to accumulate, and De Lapparent,[2] on a rather vague definition of the area of deposition, arrives at seventy-five to eighty millions, Sollas,[3] adding more than 100 per cent. to the mean rate of collection to represent the maximum, obtains twenty-six millions. But it would appear as if some moderate increase of the mean rate was the more correct procedure; for, as already pointed out, when reckoning up the beds we include both fast- and slow-forming deposits; admitting tbereby that the area in question was at one time the place of rapid growth and at another time the place of slow growth. In other words, the deposit is *not* truly wedge-shaped, but rather takes the form of a shallow basin. Hence, unless we reckoned coarse detrital rocks exclusively, we are not entitled to assume the fastest rate of deposition for the whole formation.

The foregoing method of finding the age, which has given rise to so much discussion, appears to afford its answer by a somewhat different mode of approach from that which is generally adopted. We take first the average ratio of the mass of suspended sediment to the mass of the river water.

[1] Geikie, *Brit. Assoc. Rep.*, Address to Sect. C, 1899.

[2] *Traité de Géologie*, vol. iii., p. 1860.

[3] Sollas, *Brit. Assoc. Rep.*, Address to Sect. C, 1900.

The following table contains estimates of this factor[1] :—

	Mean annual discharge in cubic feet per second.	Total annual sediment in thousands of tons.	Ratio of sediment to water by weight.
Potomac, .	20,160	5,557	1 : 3,575
Mississippi, .	610,000	406,250	1 : 1,500
Rio Grande,	1,700	3,830	1 : 291
Uruguay, .	150,000	14,782	1 : 10,000
Rhone, .	65,850	36,000	1 : 1,775
Po, . .	62,200	67,000	1 : 900
Danube, .	315,200	108,000	1 : 2,880
Nile, . .	113,000	54,000	1 : 2,050
Irrawaddy, .	475,000	291,430	1 : 1,610
Mean, .	201,468	109,650	1 : 2,731

It will be seen that the ratio of the weight of sediment to the weight of water in the cases of all the rivers cited, save two, does not differ very widely. The arithmetic mean of the ratios, as given in the table, fails to allow for the relative importance of the rivers. When we ascribe to the ratio given above for each separate river the weight proper to the discharge, the mean ratio of sediment to water is found to be 1 : 2521.

If we now take Sir J. Murray's estimate of the total annual discharge of the world's rivers, 27×10^3 cubic kilometres, we find that the total carriage of detrital sediments to the ocean is about 4·4 cubic

[1] Russell, *loc. cit.*, p. 14.

kilometres per annum. This is for the whole earth: we have now to find an approximate measure of the area of deposition. On the basis of the measurements made in the case of the Mississippi,[1] we take the area of deposition to be one-eighteenth that of denudation. To find the latter we have first to subtract from the land area of the globe the rainless areas. These are estimated as making one-fifth the total area (which is $14{\cdot}4 \times 10^7$ square kilometres), leaving $11{\cdot}5 \times 10^7$ square kilometres; and one-eighteenth of this receives the detritus, i.e., $6{\cdot}4 \times 10^6$ square kilometres. Spreading the sediment over this, we get a rate of growth of 6·9 centimetres in 100 years. This rate we increase by 50 per cent. to allow for the fact that we compute the depth of sedimentary deposits where these are most developed.

According to Sollas[2] we may reckon the geological column as about eighty kilometres. The geological age derived would be about eighty million years. The neglect of the calcareous precipitates in this computation tends to increase somewhat the estimated age.

The question we have to ask here is: Can we be four- or five-fold in error in such an estimate? In the foregoing pages we have been at pains to set forth the directions in which this method may err. But considering each weak point separately, we must admit the great improbability of failure so complete. The integral of the sediments can scarcely be 100 miles instead of fifty miles thick as estimated; much less

[1] Magee, *Am. Journ. of Science*, 1892, p. 188.

[2] Sollas, *The Age of the Earth*, p. 14.

can it be really 200 miles. Can the ratio of the areas of denudation and deposition be on the average four or five times greater than what has been measured in the case of the Mississippi? Or can the mean found from the nine rivers for the ratio of silt to water be equally far out? Or, finally, can the several rates have been so much less in the past? The answer in each case seems to be a negative, and only on the chance that several of these factors have been computed with error of the same sign, does it appear possible to reconcile the indications of the sedimentary accumulations with the storage of the products of radioactive disintegration.

The probability that such an accumulation of errors has occurred is greatly lessened by the fact that a really quite independent mode of computing the geological age gives a similar result.

The method of estimating the geological age by the accumulated results of solvent denudation requires knowledge of two quantities: the quantity of sodium discharged per annum into the ocean by all the rivers of the world, and the amount of this element in the sea. The latter is evidently the numerator and the former the denominator of the fraction, the quotient of which is the sought-for geological age.

As regards the determination of the numerator—that is the saltness of the sea—the correct value, within a few per cent., is known. Two quantities are involved in it: the average composition of sea-water and the mass of the ocean. Our knowledge of the first is favoured by the fair degree of uniformity in chemical composition prevailing in all the great seas; our knowledge of the latter by the fact that soundings are certainly sufficiently numerous to afford the average depth of the ocean: undiscovered deeps or shallows are not likely to seriously affect the volume inferred from the average depths.

In the geological application of this knowledge we are, however, exposed to the possibility that the early ocean received salts at an accelerated rate owing to the prevalence of an atmosphere possessing acid properties. It will be shown that error from this source is calculable and its limiting amount definable. Neglect of this possible correction renders our estimate of geological time proportionately too long, for it is a subtractive correction on the numerator. The reception, by the ocean, of salts dissolved directly from the coasts, or from the submerged sediments, gives rise to yet another correction which can be shown to be probably small. Its neglect again tends to an unduly great estimate of the period of denudation. The rock-salt deposits are obviously negligible as a source of error. This will appear from the fact that the rock salt in the ocean, if spread upon the dry land of the globe, would cover it to a depth of 122 metres (400 feet). The quantity of sodium now in the ocean is about $1,555 \times 10^{13}$ tonnes. The allowance to be made for the sources of error referred to we consider later.

The denominator (the sodium content of the rivers) is not so well known. It is indeed quite determinable, but we have to go on insufficient knowledge at present. We have Sir John Murray's estimate of the total amount of river water, annually discharged into the ocean, as 27,191 cubic kilometres; and his average analysis of 19 rivers, many of them principal rivers of the world. Combining these data, we find that the total amount of elemental sodium discharged yearly is $15,976 \times 10^4$ tonnes. As the 19 rivers do not represent a sufficiently large fraction of the total drainage of the continents, this figure cannot be considered final.

The most important correction on the denominator is concerned with the cyclical nature of part of the sodium.[1] For some part of it is believed with considerable probability to be wind-borne from the sea, reaching the rivers in the rain-water. The facts connected with and giving rise to this belief may be briefly stated.

(*a*) The rain falling on islands and coastal countries carries down with it a considerable proportion of chlorides and other salts, which might be referred to oceanic origin. The proportion of such salts in rain falls off as we go inland. Thus 200 miles from the coast there may be no more than $\frac{1}{12}$ of what would be observed in rain-water collected 30 or 40 miles from the sea. In India, 300 miles from the sea, a measurement showed chlorine to the extent of only 0·04 part in 100,000; whereas in British rain-water it may be from 0·1 to 21 parts per 100,000, and even more. The rate of decrease is, apparently, rapid as we proceed inland. Remote from the sea some of the chlorine in rain-water must be due to dust raised from the ground by winds: in arid, saline regions this might affect considerably the results of rain-water analyses.

[1] Joly, *Geological Magazine*, Aug., 1901, p 341.

(*b*) Coastal rivers and the rivers of small islands show abnormal percentages of chlorine compared with the sodium, indicating that but a small fraction of the latter is referable to the rocks, and the greater part in these cases to the supply from the rains.

(*c*) The chlorine in average river-water is more than the observed amount of this element in the rocks would warrant. Thus Clarke, as an approximate estimate, gives the percentage of chlorine in the igneous lithosphere as 0·07. There is apparently no measurable amount in detrital rocks; and in limestones, which constitute only some 5 per cent. of the sediments, there is very little, so that we must conclude that in the process of denudation the whole of the chlorine passes into solution. Now the sodium in igneous rocks averages (according to Clarke) 2·43 per cent., and of this it can be shown that about one-half, or a little more, goes into solution. Thus we should expect that the break-up of an average igneous rock would liberate chlorine and sodium in the ratio of 0·07 to 1·4 (say), or as 1 to 20, about. However, the rivers show a proportion of chlorine to sodium which, on the average, is 1 to 2. Hence there is about ten times as much chlorine as would be inferred from the mean chlorine content of igneous rocks, and this fact is taken as in favour of the view that much of the chlorine of rivers is wind-borne from the ocean.

This last argument must not be pushed too far. Clarke's estimate of the elements in average igneous rocks excludes extravasated or non-magmatic substances. It is based on typical rock analyses. Now chlorine, as it happens, enters with difficulty into the constitution of silicates. We find it accordingly free (or forming hydrochloric acid) in association with melted magmas in volcanoes, and rising in springs in volcanic regions. And it is found in combination with metallic minerals and in springs associated with ore deposits. The sources of supply are in these cases probably deep-seated, and would not enter into rock analyses. Possibly the slow leaching of the rocks within the hydrosphere is ultimately responsible; but, however this may be, it is evident that, so far as the river supply is concerned, we require to include all these sources. Nor must we exclude the ancient relics of congested denudation which appear in old salt lakes, bad lands, and probably in saline waters found in deep-seated bedded rocks. Again, if we go upon this argument, we may, in the case of certain rivers, arrive at the absurdity that no sodium is being denuded: for cases occur where (if the analyses are correct) we have insufficient sodium to combine with the chlorine; so that if the chlorine and sodium reach the rivers from the ocean, there can be no sodium arising from solvent effects. This, of course, is an untenable view, and demonstrates the unsoundness of the argument which gives rise to it.

The possibility must be kept in mind that chlorine is liberated from the ocean unattended by the less volatile elements. The escape of free chlorine, or, more probably, of chlorine combined with hydrogen, seems by no means improbable. In justification of such a view we have to consider the very large amount of chlorine present; and that if the chloride of hydrogen was formed by fortuitous ionic combinations near the surface—especially in agitated water—a certain percentage of the molecules should escape. The radioactive substances in the ocean will, probably, be instrumental in liberating a certain amount of free chlorine; but it is not hard to show that this contribution must be insignificant. On this view the chlorine of rain-water might only to a subordinate degree be present as sodium chloride. This, if true, would reconcile the various observations in the simplest manner.

Admitting, however, that some part of the chloride of sodium of rivers is derived from the ocean, conveyed by the winds and falling as rain, the question arises what to allow for this source of error; for evidently we cannot include wind-borne sodium in the calculation of the geological age. The only really satisfactory mode of investigating the amount of the wind-borne sodium would be by rain-water analysis. Unfortunately we possess very little knowledge on this subject. Let us proceed, however, on the assumption that the chlorine of inland rains is a clue to the cyclical sodium. The great bulk of the river-water of the world is collected at distances greater than 300 miles from the sea. If we take 0·04 per 100,000 as the proportion of chlorine in rain-water, and, making the usual allowance for concentration by evaporation before it reaches the rivers, conclude that in rivers, chlorine to the extent of 0·1 in 100,000 is wind-borne; and, further, that this chlorine enters the rivers with the same proportion of combined sodium as it would show in sea-water, we can apply the appropriate correction to Murray's estimate of river sodium.

Following Murray, we have $84 \cdot 7 \times 10^6$ tonnes of chlorine annually discharged by rivers into the ocean, and this chlorine is contained in $27 \cdot 2 \times 10^{12}$ tonnes of water. This is a proportion of about 0·3 in 100,000. The correction then amounts to taking one-third of the chlorine as coming from the ocean. We have to find what correction this is on the sodium. Now, in the ocean the chlorine is considerably in excess of the sodium, so that if all the sodium in the ocean is assumed to exist in the form of the chloride, only 82 per cent. of the chlorine is used. Hence, ocean-water contains at least 18 per cent. of its chlorine otherwise attached than to sodium: and, accordingly, what circulates from the sea to the land can only bring with it to the rivers 82 per cent. of the equivalent sodium.

One-third of the chlorine annually discharged by the rivers, on the foregoing basis, represents 15×10^6 tonnes of sodium. The true denudative supply of sodium per annum is, therefore, $131 \cdot 5 \times 10^6$ tonnes.

We have now to return to the numerator, and apply to it such corrections as may be necessary. The assumption made above, that about one-third of the chlorine in the rivers is really derived from the ocean, involves that the true, denudative supply of chlorine to the ocean by the rivers has been $56 \cdot 5 \times 10^6$ tonnes per annum; the cyclical chlorine ($28 \cdot 2 \times 10^6$ tonnes) being subtracted from the total chlorine ($84 \cdot 7 \times 10^6$ tonnes). This has been the constant rate of supply since the beginning of denudative processes, so that its integral over geological time represents what has been brought from the land. If we call the annual supply of chlorine c and the time X, the quantity is cX. Now, if we make deduction of this from the total chlorine in the ocean, we are left with a quantity of chlorine which must—for want of any alternative hypothesis—be supposed to have existed in the primeval atmosphere of the earth, and, probably combined with hydrogen, to have effected a rapid denudation of the earth-crust. If C is the existing amount of chlorine in the ocean, the primeval chlorine is $C - cX$.

Such a primeval acid denudation would undoubtedly be responsible for the comparatively rapid introduction of chlorides into the original ocean. We can make some estimate of the quantities involved. Thus, if we ascribe to the original lithosphere a composition similar to that of the average igneous lithosphere of to-day, we have (again quoting Clarke):—

Aluminium, . . .	7·90 per cent.
Iron, . . .	4·43 ,,
Calcium, . . .	3·43 ,,
Magnesium, . . .	2·40 ,,
Potassium, . . .	2·45 ,,
Sodium, . . .	2·43 ,,

We here refer only to such substances as might be supposed to enter into combination with the chlorine present. Assigning to these elements their combining proportions of chlorine, it is found that only 6·7 per cent. of the chlorine is used in bringing sodium into the primitive ocean. The total quantity of sodium thus introduced into the ocean, at the dawn of the denudative period, was

$$(C-cX)\,\frac{6{\cdot}7}{100}\times\frac{23}{35},$$

the last factor serving to convert the chlorine into its equivalent of sodium. This, when simplified, becomes $(C-cX)\ 0{\cdot}044$. The quantity of sodium which has been supplied by the rivers is found by subtracting the above amount from what is now in the ocean. If this last quantity be denoted by N, and n be the annual sodium discharge from rivers (as previously corrected), we have for the determination of geological time the equation:—

$$X=\frac{N-0{\cdot}044(C-cX)}{n}=\frac{N-0{\cdot}044C}{n-0{\cdot}044c}.$$

In this equation $N = 1{,}555 \times 10^{13}$; and from Dittmar's analysis of average sea-water, we get for the total chlorine in the ocean $28{,}769 \times 10^{12} = C$. For n we have found $131{\cdot}5 \times 10^6$, and for c, $56{\cdot}5 \times 10^6$, all in tonnes. The value of X determined from these figures is 111 millions of years. This is the period required for the accumulation of the sodium in the ocean, according to Murray's average composition of river-water and his statement of its total amount; the corrections already described being applied.

It is probable that this is the best result at present attainable according to the method of solvent denudation. It may be contended by some that a longer duration of denudation might be arrived at if we allowed for a greater circulation of ocean salts from the sea to the land. As we cannot be sure of the precise correction to apply for this source of error, it is important to show that a fixed and definite limit can be assigned to the effect of this correction upon the duration arrived at.

We have seen that in the ocean there is an excess of chlorine over sodium, so that only 82 per cent. of the chlorine can be combined with sodium. Now, let us suppose that all the chlorine in rivers is derived from the ocean, and, taking 82 per cent. of it, subtract the equivalent sodium from the average river content of that element; regarding the balance only as due to denudation. This procedure evidently constitutes a maximum allowance for the circulation of salt: an allowance obviously excessive, for we ignore the various sources of chlorine—rocks, volcanoes, springs, lodes, ancient salt deposits (not marine), etc., which must contribute some part of the chlorine, as well as the possibility that free chlorine is evolved from the ocean.

When we carry out the calculation for the geological age, proceeding as before, and making $c = 0$, we find the period of denudation to be 141 millions of years. On Murray's data this is a major limit to geological time.

The question as to whether Murray's average sodium content of river-water will be confirmed with increase of our knowledge of the subject, is a most important one. We see, also, that the ratio of the sodium to the chlorine is important, if we desire to refine upon the result obtained by treating the whole of the sodium as derived by denudation. A knowledge of average inland rain-water, so far as its sodium content is concerned, would, however, make us independent of the chlorine content of rivers. We shall probably have these data before very many years have gone over. It is of interest meanwhile to note that recent advances do not appear to suggest any very wide departure from the results for the age which Murray's figures have afforded. I am indebted to Professor Sollas for the following computation, in part founded on Clarke's collection of North American river analyses.[1]

Professor Sollas allows for the wind-borne sodium by deducting from the several rivers the sodium equivalent of all the chlorine present. The balance appears in the column headed "Sodium." We have first to recall that this is an excessive reduction, for even on the assumption that all the chlorine came from the ocean, we are only entitled to suppose that 82 per cent. of it brought in sodium. And of course even this is too great an allowance.

	Discharge per annum, cubic miles.	Sodium: tons.
St. Lawrence,	70·591	1,093,000
Androscoggin,	0·652	3,537
Merrimac,	0·911	8,187
Mohawk,	1·290	12,285
James,	2·144	16,235
Cahaba,	0·394	2,990
Mississippi,	132·355	1,322,550
Rio Grande,	0·298	33,700
Pecos,	0·083	14,640
Colorado,	5·746	1,832,000
Snoke,	1·557	74,410
Sacramento,	7·613	211,960
	223·634	4,625,494

[1] Clarke: *The Data of Geochemistry*, p. 60, *et seq.*

Dividing the total sodium by the total annual discharge of these rivers in cubic miles, we get a sodium content of 20,684 tons per cubic mile. Murray's data after deduction of the sodium chloride afford 17,501 tons of sodium per cubic mile. The new data obviously point to a diminished result when applied to the calculation of geological time.

We see, then, clearly that, going on collective facts of any authenticity, the denudation period must be less than 150 millions of years, taking a round number. And so far as we know, this is reached only by excessive assumptions. A subtractive correction of a few millions of years might be called for on the score of direct marine denudation of the coasts and shore deposits. It is not probably a large correction, for the land area exposed to denudation is, compared to the coastal area, enormous. A primeval accelerated denudation, at the time when the heat conducted to the surface of the earth was greater than now, might entail a further subtractive correction on the "age"; for, as in the case of coastal denudation, here, too, sodium is brought into the ocean otherwise than by the rivers. These are minor corrections, but they tend one way. Taking all into account, until we know more about the origin of river-borne chlorine, the computation which gave us 111 millions of years seems the best estimate we can make.

The method of determining the geological age by the sodium content of the rivers and ocean receives confirmation in the fact that moderate estimates of the mass of the detrital rocks, taken along with their average deficiency in sodium as compared with the parent igneous materials, will account for the sodium in the sea. In a previous chapter (Chapter VI) this has been referred to in another connexion. This fact brings vividly before our minds the perseverance of those denudative processes which, age by age, attended the formation of the sediments and brought the soluble substances into the ocean.

As regards the uniformitarian basis which underlies this method and the method by sedimentation, it is to be observed that there now is (and possibly always has been) an excess of land area beyond what denudation seriously affects, i.e., the riverless or rainless regions of the earth. This rainless area now amounts to about one-fifth of the entire land surface. Hence the elevation of the continents can only shift the area of denudation, and their depression cannot seriously lessen it, until it has proceeded so far that more than one-fifth the present land surface is submerged.

As to the legitimacy of prolonging into the past climatal conditions similar to those which at present prevail upon the earth, the testimony of the rocks, and the evidence to be derived from their fossil remains, are at one. Sir Archibald Geikie brings the geological testimony on this point forcibly before us in his opening address to Section C at the British Association meeting of 1899. Geologists, he points out, have found no confirmation for the view that greater denudative activity existed in the past:—"On the contrary, they have been unable to discern any indication that the rate of geological causation has ever, on the whole, greatly varied during the time which has elapsed since the deposits of the oldest stratified rocks." He instances the evidence of measured tranquillity in the deposition of the ancient Torridonian sediments, and the fact that the same tale, with endless additional details, is told all through the stratified formations down to those which to-day are in course of accumulation.

The zoological aspect of the same question is considered by Poulton,[1] and similar conclusions are arrived at. The evidence of fossil organisms seems to be incontestable. It is to be read in the proportions and structures of early woody plants and trees, in the wings of birds and insects, and in the existence of frail insect life in early times. And this evidence is greatly strengthened when the remarkable influences of severe climatal conditions, as witnessed to-day, on winged animal life, and on arboreal life, are considered. Thus while fossil insects from the Carboniferous display great powers of flight (a fossil dragon-fly showing upwards of two feet of wing-expanse), we see in storm-swept countries of the world (e.g., Kerguelen Land) the whole of the insect fauna rendered wingless. Similarly, while forests of slender Cryptogams flourished in later Palæozoic times, we find to-day that the gales on our sea-coasts practically inhibit arboreal life.

We conclude that the former physical conditions must have been much as now; and the view that stupendous rains, hurricanes, and tides then existed

[1] Poulton, Address to Section D, *Brit. Assoc. Rep.*, 1896.

is unsubstantiated by any records at our disposal. As already remarked, however, had such conditions indeed existed, an accelerated rate of denudation, and not a retarded one, must have been the result.

The work of Sir George Darwin, based on the theory of a genetic connexion between earth and moon, assigned a minor limit of fifty-six millions of years to the time since their separation. This, however, does not touch the point at issue here. Similarly we have seen how Kelvin's estimate of time since the *consistentior status* can be lengthened by the heating effects of radioactivity.

With the denudation method of estimating the age we close the case against the present indications of the radioactive method. Can the nineteen rivers upon which Murray's figures are based have given four or five times the correct sodium average? Here seems to be the only possible opening for serious error. But recent data suggest no possibility of so complete a failure of the original estimates. And this being so, what are we to conclude? That the parentage of the stored helium is not altogether to be assigned to the uranium family of elements seems to be the most natural answer, if, indeed, the migration of radioactive materials be proved an inadequate explanation. With reference to the more venerable ages arising out of the storage of lead in primitive minerals (over and above the possibilities of error), we must bear in mind the possibility that these materials may date their radioactivity back, even through many vicissitudes, to times before our denudative period began.

With an interest almost amounting to anxiety,

geologists will watch the development of researches which may result in timing the strata and the phases of evolutionary advance; and may even—going still further back—give us reason to see in the discrepancy between denudative and radioactive methods, glimpses of past æons, beyond that day of regeneration which at once ushered in our era of life, and, for all that went before, was "a sleep and a forgetting."

CHAPTER XII.

RADIUM MEASUREMENTS IN PRACTICE.

THE ELECTROSCOPE AND ITS CONSTRUCTION—RATE OF ACCUMULATION OF THE EMANATION OF RADIUM—TABLE FOR REFERENCE—EXPERIMENTAL STANDARDIZATION OF THE ELECTROSCOPE—MODE OF USING THE ELECTROSCOPE—CONDITIONS OF ACCURACY IN PREPARING SOLUTIONS—TREATMENT OF ROCKS AND OTHER SUBSTANCES—PRECAUTIONS AGAINST CONTAMINATION—MODE OF BOILING-OFF AND TRANSFERRING THE EMANATION—SOURCES OF OCCASIONAL VARIATIONS IN REPETITION EXPERIMENTS.

IN the earlier chapters sufficient has been said to enable the general theory of radium measurement to be understood. It is intended in what follows to give an account of the details of apparatus and procedure, such as will enable those having the requisite manipulative training to pursue this line of research.

The Electroscope.—The electroscope (fig. 3) is very simple in construction, and may, in fact, be made from a round glass flask of about 400 to 500 cubic centimetres' capacity, by cutting off most of the neck. A cork which has been boiled in paraffin wax is fitted to the neck and is perforated in two places. One perforation takes a glass tube leading to the drying-tube, through which gas is removed or admitted to the electroscope. The drying-tube contains phosphoric anhydride, with a close-pressed plug of cotton wool inserted between the powder and the flask, so as to avoid any possibility of the drying material being drawn into the electroscope. A glass stopcock at the outer end of the drying-tube controls the entry of gas; and for greater safety a short length of capillary tube is inserted between it and the drying-tube. This guards the electroscope from too rapid an inflow of gas. The second perforation in the cork takes a short length of glass tube, which in turn carries an inner sulphur tube of fine bore—about one half millimetre—supporting the gold-leaf system.

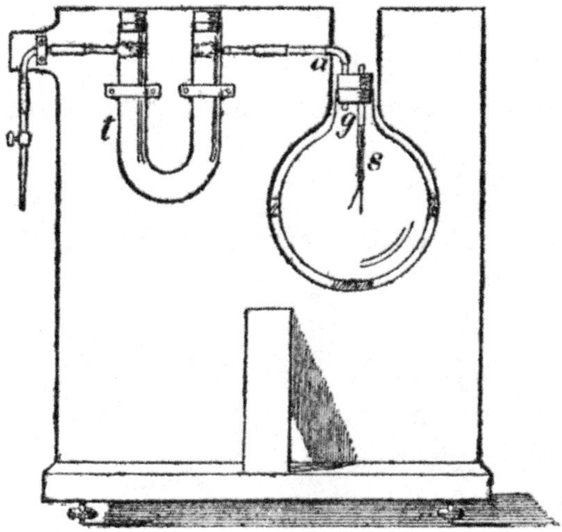

FIG. 3.

The sulphur tube is prepared as follows:—Into a length of about 30 centimetres of glass tubing, melted sulphur is drawn by air-suction, one end of the tube dipping into the vessel of sulphur. After a moment the liquid sulphur is let run back out of the tube, and is again drawn up; the process being repeated till a sufficient thickness of sulphur is deposited on the walls of the glass tube. This leaves a fine-bore sulphur tube within the glass. Letting this stand a few hours, it will be found that the sulphur has shrunk away from the glass, and if now a length of about 6 centimetres is cut from the glass tube, the tube within may be pushed forward so as to project a couple of centimetres from one end. This projecting part must not be touched by the fingers at any time. To make the sulphur tube fit air-tight in the glass, it is requisite to put a little thick Canada balsam just where it emerges from the glass. This creeps in and hardens, making a perfect joint. At this stage the glass tube supporting the sulphur tube is inserted in the cork, in such a way that the extremity of the sulphur reaches downwards to a point rather above the centre of the flask. A flat piece of brass about 2½ centimetres by 0.3 centimetre is attached while hot to the end of the sulphur tube, centrally, so that it crosses the bore of the tube. With a warm penknife the attachment of the brass strip is made air-tight in the bore of the sulphur tube.

The gold leaf is finally attached. It is a little shorter than the brass strip and about the same width, and is attached by a trace of gum just where the brass leaves the sulphur.

When the cork is finally inserted in the flask, the fit must be made perfectly air-tight, by melting the paraffin around and over the cork. It should be mentioned that in order to earth the inner wall of the flask a couple of tin-foil strips are brought in beside the cork and gummed to the interior of the flask. These strips bend down outside, meeting the glass where it is free from paraffin.

The electrification of the leaves is effected by inserting through the bore of the sulphur tube a fine platinum wire carried in a sulphur handle. This makes contact with the sealed-in end of the brass strip, and when a charged body is applied to it above, the leaves diverge. The divergence when using the instrument should not be greater than about 30 degrees. When the proper divergence is secured, the platinum wire is withdrawn.

The motion of the leaf is observed by a reading microscope, with a scale in the eyepiece, dividing the field into 100 divisions. A good stopwatch is necessary when the rates of discharge are rapid. The time required for the extremity of the leaf to move over, say, ten divisions, is noted and the rate converted to so many scale divisions per hour.

The electroscope is standardized in the following manner :—One-fifth of a gram of Joachimsthal pitchblende is carefully weighed out and dissolved in pure nitric or hydrochloric acid. The solution is made up to 200 cubic centimetres with distilled water. Thus one cubic centimetre contains one milligram of the ore. To effect the calibration, 5 cubic centimetres are measured out in a pipette into a clean flask, and about 600 cubic centimetres of distilled water added. This is boiled vigorously for twenty minutes. The flask is then closed by a good rubber stopper, and set aside for about twenty-four hours. At the end of this time the contents of the flask are again boiled, and, by a procedure to be described later, the boiled-off air and emanation are transferred in their entirety to the electroscope. During the time of closure of the flask (which must be carefully noted) the radium present in the solution, which was deprived of any original emanation by the preliminary boiling, is generating fresh emanation, and the amount of emanation generated is proportional to the quantity of radium present and to the period during which it collects.

The quantity of emanation formed depends also on the rate at which it is derived from the radium, and the rate of its own transformation. The first rate is a steady one over the time interval concerned, and we may consider that the radium produces continually n particles per second. The rate of transformation of the emanation being relatively rapid cannot be ignored. We may say[1] that when N particles are accumulated the number transforming per second is λN, where λ is the radioactive constant of the emanation or fraction transforming in one second. Obviously the number changing is variable, increasing till the equilibrium value is reached.

At any time t, however, the net rate of increase of the emanation must be $n - \lambda N$; or

$$\frac{dN}{dt} = n - \lambda N.$$

The solution of this equation is:

$$N = ae^{-\lambda t} + \frac{n}{\lambda},$$

where a is a constant.

To find the value of the constant write $t = 0$, and

[1] See Rutherford's *Radioactivity*.

then the first term on the right is equal to a; and as when $t = 0$, N must $= 0$, we have $a + \frac{n}{\lambda} = 0$, or $\frac{n}{\lambda} = - a$; also when $t = \infty$, we have the first term on the right of the equation equal to 0; and as then the maximum number, N_0, is formed, we have $\frac{n}{\lambda} = N_0$; and the equation becomes:

$$N = N_0 - N_0 e^{-\lambda t}. \qquad (1)$$

The value of λ for emanation is found by experiment to be $2{\cdot}16 \times 10^{-6}$.

We will now apply the equation to calculate the amount of emanation accumulated during, say, twenty-four hours. In this case t in seconds is 864×10^2. The numerical value of λt is therefore 0·18662.

The last member of the equation is best evaluated logarithmically. Thus if we let $x = N_0 e^{-\lambda t}$, we have

$$\log x = \log N_0 - \lambda t \log e.$$

The value of $\log e$ is 0·43429, and if we let $N_0 = 100$, the log of which is 2·00000, we find $\log x = 1{\cdot}91896$; from which $x = 83{\cdot}0$. This has now to be subtracted from 100, and finally we have from (1) $N = 17$. That is to say, there is 17 per cent. of the emanation accumulated in twenty-four hours.

The trouble of making this calculation each time we calibrate the instrument, or each time we deal with a substance which has been enclosed for less than the time required to secure the equilibrium amount of emanation, may be avoided by plotting a curve to a large scale once for all, in which the abscissae are the times since closing up the solutions, and the ordinates the percentage collected proper to the time, derived by solving the equation as above. The following table will enable such a curve to be plotted, or may be used directly for ordinary work.

Time of collection of emanation.	Percentage of equilibrium amount of emanation accumulated.
6 hours.	4·55
12 ,,	8·9
18 ,,	13·1
24 ,,	17·0
30 ,,	20·8
36 ,,	24·4
42 ,,	27·8
2 days.	31·1
2½ ,,	37·3
3 ,,	42·9
3½ ,,	47·9
4 ,,	52·6
4½ ,,	56·8
5 ,,	60·6
5½ ,,	64·2
6 ,,	67·4
6½ ,,	70·3
7 ,,	72·9
8 ,,	77·6
9 ,,	81·3
10 ,,	84·5
12 ,,	89·4
14 ,,	92·7
16 ,,	94·9
18 ,,	96·5
20 ,,	97·6

In standardizing the electroscope according to the

foregoing procedure, we, of course, require to know the amount of radium present in the standardizing solution. This we find by a chemical analysis of the uraninite or pitchblende, determining the uranium only. From this we can derive the radium, as the proportion in which the two exist in old minerals is constant. For every gram of uranium we have, according to Boltwood's latest determination, $3{\cdot}4 \times 10^{-7}$ grams of radium.[1]

The uranium is best determined by Patera's method, given in Fresenius' Quantitative Analysis (vol. ii., p. 310). From the results of the analysis, the percentage of metallic uranium is calculated. For instance, in a particular analysis, the percentage came out as 59·4. Taking this as 60 per cent. (as there is generally a slight deficiency in the amount determined by analysis), and assuming that 5 milligrams of the ore had been used in a standardizing experiment, we have, for the radium:

$$5 \times 10^{-3} \times \frac{60}{100} \times 3{\cdot}4 \times 10^{-7} = 1020 \times 10^{-12} \text{ grams.}$$

Suppose now the time of closure of this amount of radium had been 24 hours, then the quantity of radium represented in the increased rate of discharge of the electroscope would be

$$\frac{17}{100} \times 1020 \times 10^{-12} = 173 \times 10^{-12}.$$

This, in fact, is the quantity of radium which would produce the observed effect on the electroscope, if the emanation in equilibrium with it had been collected.

[1] Boltwood, *Am. Journ. Science*, June, 1908, p. 504.

We now see that the calibration is easily completed. For suppose the normal leak of the instrument is 15 scale-divisions per hour, and the leak, three hours after the introduction of the emanation, was 335 scale-divisions per hour, then the gain has been 320, and this is due to the emanation in equilibrium with 173×10^{-12} grams of radium. The constant of the electroscope is found from this by calculating how much radium would give rise to an increased rate of discharge of 1 scale-division per hour: that is, we divide the quantity of radium by 320—the final result being that the constant is $0{\cdot}54 \times 10^{-12}$.

When now in future experiments the gold leaf shows an increased rate of collapse, we have only to multiply by this constant, and we get the radium responsible. For example, if the increased rate was only 40 divisions, then the radium present in the solution was $21{\cdot}6 \times 10^{-12}$, and if 10 grams of rock had been contained in the flask, we enter as the radio-activity of the rock, $2{\cdot}2 \times 10^{-12}$.

This mode of using the electroscope renders the method entirely comparative; and when the analysis of the standardizing substance—preferably pitchblende (uraninite)—has been made, we are in no doubt as to the indications of the instrument. It will be found that successive standardizing experiments give sensibly the same constant, even when the quantities of emanation introduced are varied largely. This certainly holds up to, and probably over, 800 scale-divisions per hour rate of discharge. If the electroscope is to keep its normal rate of discharge constant, there must be careful attention to the exhaustion of the instrument before and after each experiment: the electroscope in each case being refilled with air. Occasional renewal of the drying material is of course required.

Preparation of Materials and experimental Procedure.—The extraction of the emanation from a solution and its transference to the electroscope may now be considered.

The conditions conducive to accuracy in these operations are: (1) The preparation of a solution free from any trace of precipitate, or accidental contamination with radioactive matter; (2) Thorough boiling of the solution when extracting the emanation, the ebullition being of the effervescent type (*i.e.*, with very numerous small bubbles); (3) Complete transference to the electroscope of the boiled-out gases and intermixed emanation.

To successfully attain the first condition is in the case of certain rocks a matter of considerable difficulty. The mode of preparation of a solution is as follows:—The rock is ground in the agate mortar till it passes a sieve of 120 meshes to the inch. Ten to fifteen grams are then weighed out, and mixed with the anhydrous carbonates of sodium and potassium (taken in equal proportions); the proportion of the carbonates to the rock varying from twice to two and a-half times the weight. The whole, being well mixed, is fused in a platinum crucible having a capacity of not less than 90 c.cs. In this process the blowpipe flame must not play into the crucible. The first part of the fusion may with advantage be carried out with a strong bunsen, and, after twenty minutes' treatment, the foot-bellows and blowpipe should be brought into operation. When effervescence ceases, the contents of the white-hot crucible are tilted into a *dry* platinum dish. The hardened melt, collected from the crucible and the dish, is then crushed to fine fragments, thrown into distilled water (400 c.cs.), and left to digest on the water-bath; the beaker containing the solution being kept covered with a clock glass.

After some hours if the melt is not broken up in the water, the hard parts are separated and crushed in a biscuit-ware mortar and again returned to the beaker. Finally the contents of the beaker are filtered, and the filtrate, which must be quite limpid, transferred to a flask and closed as the alkaline solution. The residue is now stirred up in a little distilled water and treated with about 40 to 60 c.cs. of pure hydrochloric acid which has been distilled in the laboratory over chloride of sodium. The re-distillation of the acid is necessitated by the presence of radium in most "pure" HCl on the market, and, what is a source of error in the opposite direction, the presence of sulphuric acid. If the rock is of the acid class (that is, has a high percentage of silica), the solution with the acid will in most cases be complete, and the liquid may at once be closed in a flask as the acid solution. The two flasks are then set aside, being numbered in correspondence with the laboratory notes, and dated. It is necessary to leave an air space in the flasks, such as will allow of subsequent ebullition, but no more—about 100 c.cs. The flasks should be of about 600 c.cs. capacity.

If the action of the acid results in an undissolved

residue, this must be carefully filtered off, dried on the filter paper, which is then ignited, and the ash and residue re-fused in the platinum crucible with a small quantity of the mixed carbonates. It will now be found, in nearly all cases, that the melt goes into complete solution when leached with hot distilled water and treated with a few c.cs. of HCl. The solution is added to the contents of the "acid" flask.

In this process it will be seen that a much less quantity of chemicals is used to effect decomposition than is ordinarily recommended; it will be found, however, that the quantities stated are quite adequate. It is important to keep down the amount of chemicals, for not only will the carbonates certainly contain some radium, but the presence of much dissolved matter hinders the ebullition and the free evolution of the emanation. Tests must be applied to the carbonates by dissolving and closing up quantities of about 100 grams, and then boiling off any emanation, etc., as in the ordinary experiment. The correction for radium may amount to about $0{\cdot}13 \times 10^{-12}$ grams per gram of mixed carbonates.

It sometimes happens that a solution put by develops a faint turbidity after some days. Often this cannot be filtered out, and if the solution is retained for experiment, it must be noted that the resulting reading is probably only a minor value, unless the precipitate is observed to go into complete solution on ebullition. It is, indeed, very generally the case that we are not sure of obtaining more than a minor limit to the true quantity of radium present. It is easy to understand the reason for this. The total amount of emanation developed in the solution is extremely small in volume. The recent experiments of Rutherford[1] have shown that the volume of the emanation in equilibrium with one gram of radium is 0·585 cubic millimetre. It results that in a solution of ordinary radioactivity there are only a few billionths of a cubic millimetre of the gas when in equilibrium amount. Now, if any cause leads to the precipitation of the radium, it is certain that the true amount is not boiled off, no matter how prolonged the ebullition. A trace of sulphuric acid appears to lead to this result, provided any barium is also present in solution; there are comparatively few rocks without it. This result can be demonstrated experimentally. It will be remembered that radium is an element of the barium group, and that its sulphate is at least equally insoluble. From this it will be seen how important is the attainment of limpidity in the solutions. It may be mentioned in this connexion that it is not uncommon for a second experiment on the same solution to show a diminished yield of emanation if a precipitate has developed since the first experiment was made. The really astonishing fact is the removal of so small an amount under any circumstances.

Some of the materials dealt with in research are, of course, much more easily prepared than siliceous rocks. Carbonates, manganese nodules, haematite, gypsum, etc., may often be brought into solution by pulverization

[1] Rutherford, *Proc. R. S.*, A, vol. 81, 1908, p. 162.

and heating with hydrochloric acid, and may then be diluted and bottled without further trouble. Residues, in such cases, however, always involve risk of error unless there is reason to believe that they are purely siliceous. Sea waters and river waters are preferably, but not necessarily, concentrated by evaporation. In all cases they should be acidified by hydrochloric acid.

Regarding the precautions to be taken against the introduction of radium contamination, the test of performing a blank experiment must not be omitted; that is, the whole procedure must be gone through with the chemicals only. Again, the experiments are not reliable if performed in a room wherein radioactive preparations have previously been dealt with at any time. The investigator must not himself be concerned with radium preparations while carrying out these delicate measurements; and if he has been so engaged at any time in the past, he must test (by placing them in close proximity to a suitable electroscope, or bringing them near to a sensitive screen of phosphorescent zinc sulphide) any articles likely to retain traces of adhering radium, as, for instance, the handles of the laboratory doors, his watch, keys, or other articles carried in his pockets. Further, it is necessary to discard the use of any apparatus formerly in use in connexion with radioactive salts.

The attainment of the second and third conditions of accuracy, i.e., the free ebullition and perfect transfer to the electroscope, requires special arrangements.

In fig. 4 (p. 266) the receiver *R* is of about 500 cubic centimetres capacity. It has two tubulures. During ebullition it is placed as shown, in connexion with the flask containing the solution; a glass condenser, in which cold water circulates, intervening. The attachment of the flask to the condenser must be rapidly performed. Before making up these connexions the

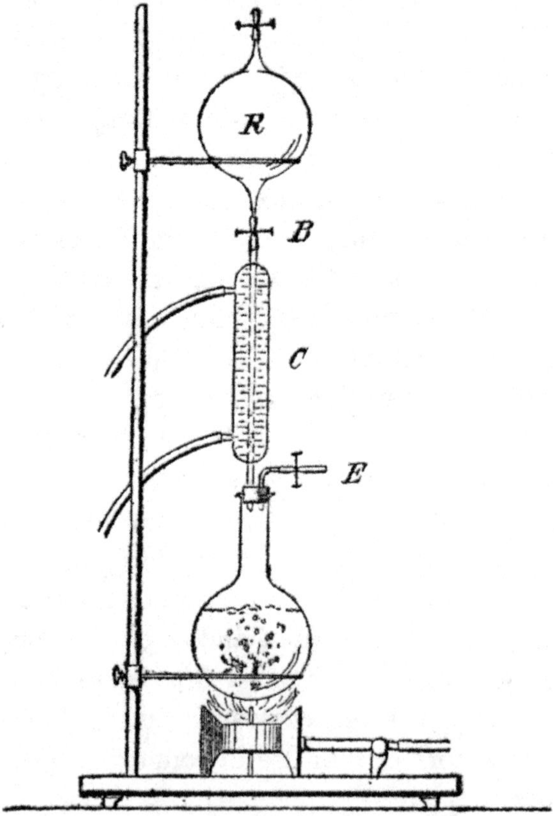

FIG. 4.

receiver is exhausted of air; thus when, after the connexions are made, the pinch-cock B is opened, the pressure falls in the flask. This condition both enables ebullition to be maintained without risk in the closed system of vessels, so long as the water circulates in the condenser, and facilitates the withdrawal of the gases dissolved in the solution. The tubing forming the connexions must be thick, soft rubber, and seamless.

Ebullition is carried on for half an hour, and the effervescent type of boiling is secured very simply by putting a small trace of powdered steatite into the solution. This is best done by dropping the powder, before making the connexions, into the tube of the condenser, where it adheres till the condensed steam, dripping back into the flask, carries it down into the solution. There is very little radium to be expected in this substance; for safety it should be tested, however.

When ebullition has proceeded vigorously for about thirty minutes, the burner is extinguished, and cold distilled water is let flow in at the tube *E* which passes through the rubber stopper fitting the condenser to the flask; this tube being kept closed, throughout the boiling, by a pinch-cock. The distilled water is let fill the air space in the flask and rise into the lower tubulure of the receiver above. Thus the entire quantity of boiled-out gas is driven into the receiver. The pinch-cock *B* is then closed on the rubber connexion with the condenser, and the receiver removed to the electroscope. The air being pumped out of the latter, the receiver is attached in such a manner that on opening the connexions the gaseous contents are drawn into the electroscope, while water rises and fills the receiver. Thus a complete transfer is effected. An important feature of this arrangement is that the emanation is as short a time as possible in contact with water in which it may be absorbed. Its solubility is known to be as much as one-twentieth that of carbon dioxide; considerably exceeding that of oxygen.

Before the transfer it is advisable to keep the gold leaf for some time under observation, to make sure that the leak is normal. About ten minutes after the transfer, a reading of the rate of discharge should be taken, and, as before stated, again in three hours. A comparison of the readings assures us that we are dealing with the emanation of radium, and eliminates any possibility that a spurious effect may obtain. The leak will be increased about 30 per cent. in the three hours.

When the material under investigation has to be divided between an acid and an alkaline solution, the separate examination of each solution may be simultaneously carried out by duplicating the apparatus shown in figure 4. Two electroscopes should be kept in readiness for dealing with the emanation withdrawn. In this way two solutions may be examined with hardly more expenditure of time and labour than would be required for one.

It sometimes will be found that successive results with certain rocks, even from the same hand specimen are not consistent. This may arise from failure to effect complete extraction of the radium in one of the experiments, or—and if the solutions are well prepared, this is the most probable explanation—it may be due to a real want of homogeneity in the radium distribution in the rock. In an exaggerated way we experience in such cases all the difficulties which beset the assay of precious metal in a rock in which the ore is sporadically distributed; exaggerated because an invisibly minute aggregate of a uranium ore can exert a considerable influence on the experimental result. It is suggestive that cases of inconsistency are seldom met with in sedimentary rocks which are not highly metamorphosed, but retain their original state of subdivision. It is probable that in sediments derived by precipitation there is very perfect distribution of the uranium. This is borne out by the consistency of the results on deep-sea sediments, even in the case of materials of the same character from widely separated parts of the ocean. In recent lavas, again, a close consistency in successive experiments is generally observed.

The pleochroic haloes, which have been described in Chapter III, provide us with ocular demonstration of the sporadic distribution and segregation of the uranium in certain ancient rocks.

APPENDIX A.

I. Rise of temperature in a radioactive layer of thickness D, losing heat only at the surface, and not receiving heat from below. The surface is supposed to be maintained at a constant temperature from which the temperatures in the layer are reckoned. The layer has existed so long that there is equilibrium between the radioactive supply of heat and the surface loss.

Let $Q + \frac{dQ}{dx}dx$ units of heat enter at lower surface of an elementary volume having the thickness dx, and at the distance x from the surface, and let Q units emerge at its upper surface in unit time; then

$$-\frac{dQ}{dx}dx = qhAdx,$$

where q is the radium-content of unit volume, and h the heat developed per second per gram of radium; A is the area. Hence

$$-\frac{dQ}{dx} = qhA.$$

Also $Q = AK\frac{d\theta}{dx}$, where $\frac{d\theta}{dx}$ is the gradient at the distance x from the surface, and K the conductivity; and

therefore $$\frac{dQ}{dx} = AK\frac{d^2\theta}{dx^2},$$

and $$K\frac{d^2\theta}{dx^2} = -qh.$$

Integrating, $$K\frac{d\theta}{dx} = -qhx + C_1.$$

Now, $\frac{d\theta}{d} = 0$ when $x = D$, i.e. at base of layer, therefore $C_1 = qhD$.

Hence $$K\frac{d\theta}{dx} = -qhx + qhD = qh(D - x).$$

Integrating again,

$$K\theta = qhx\left(D - \frac{x}{2}\right) + C_2;$$

but when $x = 0$, $\theta = 0$, therefore $C_2 = 0$.

Therefore, $$\theta = \frac{qhx}{K}\left(D - \frac{x}{2}\right), \qquad (1)$$

and when $x = D$,

$$\theta = \frac{qh}{2K}D^2. \qquad (2)$$

II. Rise of temperature in a sphere of radioactive materials, of radius R, which has existed so long that there is equilibrium between the heat generated within the sphere and the heat escaping at the surface. The surface is supposed to be maintained at a constant temperature from which temperatures within are reckoned.

Let Q units of heat enter in unit time a concentric spherical layer of radius r, and of thickness dr; and let $Q + \frac{dQ}{dr}dr$ units emerge.

$$\frac{dQ}{dr}dr = 4\pi r^2 dr \,.\, q \,.\, h,$$

where q and h have the same significance as in (1).

Also,

$$Q = -KA\frac{d\theta}{dr} = -K \,.\, 4\pi r^2 \,.\, \frac{d\theta}{dr},$$

where $\frac{d\theta}{dr}$ is the gradient at the distance r from the centre.

$$\therefore \quad \frac{dQ}{dr} = \frac{d\left(-K \cdot 4\pi r^2 \frac{d\theta}{dr}\right)}{dr} = 4\pi r^2 qh,$$

$$K \frac{d}{dr}\left(r^2 \frac{d\theta}{dr}\right) = -qhr^2.$$

Integrating,

$$Kr^2 \frac{d\theta}{dr} = -qh\frac{r^3}{3} + C_1$$

but when $r = 0, \ \frac{d\theta}{dr} = 0; \quad \therefore \ C_1 = 0,$

and $$K\frac{d\theta}{dr} = -qh\frac{r}{3}.$$

Integrating again,

$$K\theta = -qh\frac{r^2}{6} + C_2;$$

but $\theta = 0$ when $r = R$;

$$\therefore \quad C_2 = \frac{qhR^2}{6},$$

and $$\theta = \frac{qh}{6K}(R^2 - r^2). \qquad (3)$$

APPENDIX B.

Geophysical and Radioactive Data.

The Earth.

Mean radius, ...	$6{\cdot}37096 \times 10^8$ cms.
Volume, ...	$1{\cdot}083 \times 10^{27}$ cub. cms.
Surface, ...	51×10^{17} sq. cms.
Mass, ...	$5{\cdot}987 \times 10^{27}$ grams.
Mean density, ...	5·527.

The Ocean, Land and Rivers.

Area of land on the globe, . .	$14{,}456 \times 10^4$ sq. kiloms.
Ratio of land to water, . .	1 : 2·54 (Wagner[1]).
Area of the ocean, . . .	$36{,}718 \times 10^4$ sq. kiloms.
Mean depth, ...	3·851 kiloms. (Murray[2]).
Volume, ...	$1{,}414 \times 10^6$ cub. kiloms.
Mass, ...	$1{,}452 \times 10^{15}$ tonnes.
Density of average sea-water, .	1·027.
Mean radium content of sea-water,	$0{\cdot}017 \times 10^{-12}$ grams per cub. cm. (J.).
Total salts in the ocean (if 3·5 °/₀),	508×10^{14} tonnes.
Mass of sodium, ...	$1{,}555 \times 10^{13}$ tonnes.
Mass of chlorine, ...	$2{,}877 \times 10^{13}$ tonnes.

Annual river discharge, 27,191 cub. kiloms. (Murray[3]), or 27×10^{12} tonnes, nearly.

[1] Wagner, *Scottish Geographical Mag.*, 1895, p. 185.

[2] Sir J. Murray, *ibid.*, 1888, p. 1.

[3] *Ibid.*, 1887, p. 76.

Sodium carried annually by rivers to ocean, $15,976 \times 10^4$ tonnes (Murray[1]).

Chlorine carried annually by rivers to ocean, 847×10^5 tonnes (Murray[1]).

The Rocks and Sediments.

Mass of igneous rock, parent to sodium in the ocean and to sediments, 833×10^{15} tonnes (J.).

Mass of detrital sediments derived from parent rock, 558×10^{15} tonnes (J.).

Mass of precipitated sediments on ocean floor, 195×10^{15} tonnes (J.).

Estimated mass of calcareous rocks on the land, 29×10^{15} tonnes (p. 58).

Mean radium content of igneous rocks, $5{\cdot}5 \times 10^{-12}$ grams per gram (J.).

Mean radium content of sedimentary rocks, $4{\cdot}3 \times 10^{-12}$ grams per gram (J.).

Average sodium content of igneous rocks, 2·43 °/ₒ (Clarke).

Average sodium content of sedimentary rocks, 0·9 to 1°/ₒ (Clarke).

Thermal and Radioactive Constants.

Heat escaping from the earth per second per sq. cm. on gradient of 32 metres to 1° C., 125×10^{-8} gram degrees.

Ditto, per annum for whole earth, 2×10^{20} gram degrees.

Atomic weight of uranium, 238·8.

Atomic weight of radium, 226·4.

Half period of uranium, 5×10^9 years.

Fraction of uranium transforming each year (λ), $1{\cdot}4 \times 10^{-10}$,

Half period of radium, 1760 years.

Life of radium, 2540 years.

[1] Sir J. Murray, *Scottish Geographical Mag.*, 1887, p. 76.

Fraction of radium transforming each year (λ), $3{\cdot}94 \times 10^{-4}$.

Fraction of radium emanation transforming each second (λ), $2{\cdot}16 \times 10^{-6}$.

Quantity of radium in equilibrium with 1 gram of uranium, $3{\cdot}4 \times 10^{-7}$ (Boltwood).

Helium generated per year per gram of radium in equilibrium with the related elements, 316 cub. mms. (Rutherford).

Number of α-particles emitted per second per gram of radium itself, $3{\cdot}4 \times 10^{10}$ (Rutherford).

Volume of emanation in equilibrium with 1 gram of radium, 0·585 cub. mms. (Rutherford).

Number of ions created by an α-particle of radium, about 10^5.

Charge carried by an α-particle, $9{\cdot}3 \times 10^{-10}$ E. S. Unit. (Rutherford).

Mass of an α-particle, $6{\cdot}18 \times 10^{-24}$ grams (Rutherford).

Average velocity of an α-particle from radium in equilibrium with its products of rapid change, $1{\cdot}77 \times 10^9$ cms. per sec. (Rutherford).

Heat given out per second per gram of radium in equilibrium with the related elements, $5{\cdot}6 \times 10^{-2}$ gram degrees.

Ditto, in one year, $176{\cdot}4 \times 10^4$ gram degrees.

Total heat given out during radioactive transformation of 1 gram of uranium (about), $4{\cdot}48 \times 10^9$ gram degrees.

The radio-thermal constants are deduced as follows:—

1. A standard preparation of radium showed by direct experiment 110 gram degrees per gram per hour = 0·0306 gram degrees per second. This involves the activities of radium, emanation, radium A, radium B, and radium C.

Now, Boltwood (*Am. Journ. Science*, April, 1908) found

the relative ionizing activities of the several members of the uranium family to be as follows:—

Uranium, ...	1·00
Ionium, ...	0·34
Radium, ...	0·45
Emanation, ...	0·62
Radium A, ...	0·54
" B, ...	0·04 (?)
" C, ...	0·91
" F, ...	0·46
Actinium products, . . .	0·28
	$4·64 \times$ uranium.

Direct experiments on uranium minerals afforded the nearly identical result, $4·69 \times$ uranium.

The sum of the activities involved in the heat measurement is 2·56. Hence the total heat will be found from the proportion

$$x : ·0306 :: 4·64 : 2·56,$$

from which

$$x = 5·5 \times 10^{-2}.$$

The value $4·69 \times$ uranium gives $x = 5·6 \times 10^{-2}$ gram degrees per second per gram of radium.

2. By measurement of the electrostatic deflexion, and of the magnetic deflexion of the α-particles from radium C, Rutherford (*Phil. Mag.*, Oct., 1906), determined their initial velocity to be $2·06 \times 10^9$ cms. per second. The mass of an α particle is 3·84 times that of an atom of hydrogen, or $6·18 \times 10^{-24}$ gram (Rutherford, *Proc. R. S.*, vol. 81 A, 1908, p. 172). Hence,

$$\tfrac{1}{2}MV^2 = 13·1 \times 10^{-6} \text{ ergs};$$

or, as $3·4 \times 10^{10}$ particles are expelled per second by the radium C in equilibrium with 1 gram of radium, the energy developed per second is $44·54 \times 10^4$ ergs.

Referring now to Boltwood's table given above, we find that the activity of radium C is to the total activity of the uranium series as 0·91 is to 4·64. Hence the total energy would be $227{\cdot}1 \times 10^4$ ergs. Dividing by $4{\cdot}184 \times 10^7$ (the mechanical equivalent of 1 gram degree), we have the total heat developed per second per gram of radium $= 5{\cdot}4 \times 10^{-2}$ gram degrees.

This last method takes no account of the energy of recoil of the atoms emitting α particles. This adds about 2 per cent. The β and γ radiation are also neglected.

3. According to Von Schweidler and Hess (*Le Radium*, Feb., 1909) the evolution of heat per hour is 118 gram degrees per gram of radium. This is $5{\cdot}95 \times 10^{-2}$ calories per second, or $187{\cdot}5 \times 10^4$ per annum, per gram of radium in equilibrium.

4. The direct determination, by H. H. Poole, of the heat evolved by Joachimsthal Pitchblende, will give a higher value than that which is deduced above.

INDEX.

Printed in Dunstable, United Kingdom

91221356R00178